T0313186

The Psychology of Lean Improvements

Why Organizations Must
Overcome Resistance and
Change the Culture

The Psychology of Lean Improvements

Why Organizations Must Overcome Resistance and Change the Culture

Chris A. Ortiz

CRC Press
Taylor & Francis Group
Boca Raton London New York

CRC Press is an imprint of the
Taylor & Francis Group, an **informa** business

A PRODUCTIVITY PRESS BOOK

CRC Press
Taylor & Francis Group
6000 Broken Sound Parkway NW, Suite 300
Boca Raton, FL 33487-2742

First issued in hardback 2019

© 2012 by Taylor & Francis Group, LLC
CRC Press is an imprint of Taylor & Francis Group, an Informa business

No claim to original U.S. Government works

ISBN-13: 978-1-4398-7879-8 (hbk)

Visit the Taylor & Francis Web site at
http://www.taylorandfrancis.com

and the CRC Press Web site at
http://www.crcpress.com

Contents

Introduction

I want you to think for a moment about your company. Regardless of if you are the owner, the president, an upper manager, or farther down the ranks, what is it that you cherish most about the organization? What do you hold dear to you? Is it the generation of revenue, its market share, the office or factory environment, or the people? Now think of all the elements of running a successful business. You may work in manufacturing, warehousing, an office, food processing, an auto collision repair shop, a bank, or a college; it really does not matter. A business is made up of a collection of people and processes.

From my perspective, people are truly the lifeblood of any company. Of course, there are the customers, the vendors, the products or services, and the processes that are needed to perform the work. So, it is safe to say that all of these items require well-thought-out investments in their future so the company can grow and prosper. As a business owner myself, my employees are the most important asset I have and they are what I cherish the most. The manner in which we operate as a team must be smart, healthy, professional, and focused to ensure success for our clients. So, it is equally important that the environment in which we all work is a place that embraces continuous improvement. It is the business environment or process that creates the behaviors of the culture, and this environment must be continually evaluated for improvement.

To get a better understanding of the people and their thinking, let's first define what "psychology" is because I am by no means a psychologist. According to Merriam-Webster.com, psychology is defined as "the study of mind and behavior in relation to a particular field of knowledge or activity." This book can be classified as how human functions and behaviors react to inefficiency and the fixing of dysfunction as it relates to business. I have observed what happens to people's thinking when it comes to Lean or any other improvement endeavor. Regardless, I am not a trained, certified, or licensed practitioner of psychology. I am a Lean implementer, pure and simple.

Lean is a business improvement model and philosophy of waste reduction that enables a company to reduce cost, improve quality, and provide faster delivery for its customers. It is an all-encompassing business strategy of continually improving the people and processes of an organization.

I have seen firsthand not only resistance to continuous improvement, but also what happens to the behavior of leaders and their employees during Lean transformation. There is a huge psychological battle that goes on in these transformations. Businesses cling to the eight deadly wastes of Lean and find ways to place continuous improvement on the back burner. Whether through poor leadership, an inability to change, poor investment in the company, or pure ego, this commitment to inefficiency is quite perplexing.

Waste exists at every level of the organization and it has a profound impact on the thinking of your people. Every company has waste, not just manufacturers. It is like a virus that really just appears and spreads as time goes on in a poorly run company. No one individual is responsible for its creation; it just happens. To help better prepare those of you new to Lean and waste, below is a quick breakdown of the eight deadly wastes:

Overproduction is the act of making more product than necessary, completing it faster than necessary, and before it is needed. Overproduced product takes up floor space, requires handling and storage, and could result in potential quality problems if the lot was made wrong. Offices can overproduce work orders or general paperwork, working on the wrong folder and creating piles of information not ready for anyone. Overproduction has the biggest effect on the brain and will be broken down in Chapter 3.

Overprocessing is the practice of taking extra steps, rechecking, reverifying, and overperforming the work. This often occurs in fabrication departments when sanding, deburring, cleaning, or polishing is done too long because the required finished state is not visible. Machines also can overprocess when they are not maintained properly. Overprocessing is very common in administrative processes where information errors are not unusual, and the employees are required to triple-check everything.

Waiting occurs when important information, tools, and supplies are not readily available, causing machines and people to be idle. Imbalances in workloads and cycle times between processes can cause waiting as well.

Motion is the movement of people, often when they are looking for things. Where are my tools? Where are my parts? Where is the work instruction? Where is the supplier folder?

Transportation is the movement of parts, paperwork, and product throughout the facility or office. Often requiring a forklift, hand truck, or pallet jack, transportation occurs when consuming processes are far away from each other and not visible. Inefficient office and factory layouts breed lots of transportation.

Inventory is a waste when manufacturers tie up too much money, and hold excessive levels of raw, work-in-process, supplies, and finished goods inventory.

Defects are any quality deficiency that causes scrap, warranty claims, and rework hours as a result of mistakes made in the factory or office. Often some of the biggest mistakes made are before the product is even built. Also, banks, hospitals, colleges, and many other pure office environments create a multitude of mistakes that affect the internal operations and the end customer.

Wasted human potential is the act of not properly utilizing employees to the best of their abilities. People can only be as successful as the limits of the process they are given to work in. If a process inherently has motion, transportation, overprocessing, overproduction, periods of waiting, and defect creation, then that is exactly what they will do. That is wasted human potential.

The Psychology of Lean Improvements is a unique look at the mentality behind avoiding Lean transformations and why businesses are more prone to holding on to waste and inefficient processes. Almost like a commitment to inefficiency, there is an inherent fear of continuous improvement in today's businesses, and this reluctance toward a Lean journey is becoming self-destructive. The book also will break down the fear of change within executives and other organizational leaders.

The book discusses how companies have forgotten who their customers are, and what I call the *psychology of dysfunction*. The psychology of dysfunction is the odd hold that employees have on excessive inventory, overproduction, poorly designed factories, ineffective business strategies, the concept of staying busy, and living with waste on a daily basis.

Lean is a battlefield, and any of you reading this book and leading Lean journeys understand precisely to what I am referring. In the early stages of a Lean journey, organizational leaders are initially placed into a corner to first admit that there is a need to discard old ways of working and change how they operate. Regardless of what the financial statements may say, as the world around them begins to change, leaders must adapt and change with it. The fear of change that exists in leaders is the feeling that major organizational adjustments will need to be made to how they work, the processes that make product or services, the people, and the overall thinking to react to the needs of customers and the market.

The business improvement tool of Lean is powerful and sometimes so powerful that even the strongest and most stable leaders avoid its implementation. The power of Lean can be frightening to some. In any new business endeavor that is to change business strategy and culture, there are temporary dips in performance that may occur while adjusting to the new way of working. In a manufacturing environment, a new production line may be set up to reduce work-in-process (WIP) levels down to single-piece flow. Structured standard work is needed to make this transformation and during the transition from WIP to single-piece flow, productivity and output may drop as the production workers learn their new environment. This short-term decline in overall metrics is frightening to company leaders, especially within factories that cannot afford this "hiccup." Usually this is an indication of business dysfunction anyway. It is the thought of change or the fear of change that stops the Lean journey from starting. Money is being poured out of the business, bled away due to the inability to admit that there is room to improve combined with the continuation of poor business practices, all because of this fear of change.

Chapter 1 (The Psychology of Change) takes a closer look at the reasons businesses are lacking in the creation of visions for growth and prosperity and how their hold on waste is not only odd, but potentially catastrophic to their companies, their industries, and the economy. Chapter 1 breaks down this fear of change at the top level and provides solutions and guidelines to help alleviate this fear so continuous improvement may begin.

Chapter 2 (Leading the Lean Journey) describes how some organizations have forgotten who the customer is and that now is the time for Lean leaders. Businesses sometimes are not set up to accommodate the needs of the customers and then wonder why they cannot deliver their products on time, or perform their services in a way that satisfies their needs. Becoming a Lean leader and changing your current thinking about business will be all about your perception of your customers. What is the customer willing to pay for? Are your processes and infrastructures in place to provide the most competitive cost, quality, and delivery model they want? And, is your business in a position to change when it is necessary to change? Regardless of your company, deciding to "go Lean" is a bold move, and Lean provides tools and techniques to adapt and change as needed. As a Lean leader, showing your ability to change and support change all the way down to the frontline workers will promote a strong bond within the organization. Employees look up to proactive and forward-thinking leaders. If you as a leader are not willing to change when circumstances demand it, you cannot expect your people to do it either. Change won't happen.

During economic downturns, organizations will hold off on improving their businesses and go into survival mode. Often the best time for business improvements is when the economy has slowed. It always feels more appropriate to invest money and time into Lean when revenue is flowing smoothly. This is an equally ripe time to invest in innovation and growth, but evaluating current business practices when times are rough can be very smart. This chapter explains how to successfully embark on Lean business practices during this downturn. It is the best time to dissect your organization and prepare it for brighter times to come.

You will find Chapter 3 (The Psychology of Waste) to be extremely interesting as I explain an organizational culture's attachment to the eight deadly wastes, general dysfunction, and clutter. I vigorously challenge the Economic Order Quantity (EOQ) model to which many companies adhere and adopted from business school. One of the eight deadly wastes of a business is excessive inventory. This inventory can come in many forms. In manufacturing, it is the excessive purchasing of raw material, supplies, and parts. In office environments, it is overbuying of office supplies and general "hoarding" of items. Part of the EOQ model is to keep unit costs to a minimum, allowing for more profitability per unit sold—a unit being a product. There is an illusion behind the "discount" from purchasing larger quantities. However, keeping in mind that the EOQ model works well in many cases, it can be very costly in the long term for when it does not. I provide the psychology behind this overbuying and the perceived notion of savings.

This chapter also discusses the concept of "staying busy" and the need to feel overworked. The waste of overproduction is discussed as well. In manufacturing, overproduction is the most costly waste that exists because it drains labor dollars and inventory. It creates significant bottlenecks in the production processes, potentially can create quality problems, drives inventory costs, underutilizes people, and supports the EOQ model of bulk buying. Overproduction is, by far, the most abused waste. Ironically, it is the waste that seems to "feel good," and it has a unique psychological effect on people. The extra inventory that has accumulated creates a false sense of accomplishment. Whether it is piles of product or partially completed product, work orders, general paperwork in an office, or any form of overproduction, there is a feeling of efficiency that in actuality is an indication of a dysfunctional process.

Chapter 4 (The Psychology of Dysfunction) takes a unique look at inefficiency. When organizations avoid improving their operations, the culture comes up with unique ways of dealing with the dysfunction. I have seen hundreds of manufacturing and administrative processes where people, who work there day in day out, will find a way to simply make things work. From makeshift workstations to the unsafe modified cart to the extra checking and verifying to bringing personal tools to work, etc., workers make waste work and, in a lot of cases, streamline the waste. Waste also creates a misconception of working hard.

People are only as successful as the process in which they are given to work. If the process they work in is conducive to quality errors, lots of walking, moving products everywhere, working on the working things at the wrong time, extra checking and redundant steps, well, guess what they are going to do all day? It's not brain surgery. No wonder they feel busy—because they are. However, the illusion is that they are not doing the work the customer cares about. If improving the process is not the option, then human psyche is to find a way to make it work. Of course, nothing has been improved; the wastes have only been dealt with. Then, in the end, we wonder why we are missing deliveries, made mistakes, spent more money, and ultimately lost a customer. Chapter 4 holds nothing back on the psychology of dysfunction.

Solutions and answers begin to appear heavily in Chapter 5 (Making Change Happen with 5S). The first four chapters are more of a detoxification of the brain. Chapter 5 is written to help you resolve these psychological changes that are going on. One of the key elements of making change is how the culture perceives the actual physical working environment. 5S (sort, set in order, shine, standardize, sustain) is a powerful improvement tool with which a highly visual and organized workplace is created. People have the ability to adapt to the changes around them. Although these changes will create a certain level of resistance, if the physical environment is changed for the right reasons and improves performance, then it is only natural to adapt. For instance, most of you reading this book drive a car or at least commute on some form of public transportation. During daily travel, we become accustomed to certain routes and routines. There may be certain roads you always take to work. You may stop at a favorite coffee shop every Monday through Friday to get your morning beverage. Based on experience from living in the same location for awhile, you are aware of certain roads and routes to not take due to heavy traffic or length of trip. The roads we follow also have consistent visual guidelines such as stoplights, yield signs, stop signs, crosswalks, etc. We are aware of them and, for the most part, adhere to the rules and restrictions of the road as we travel to and from work. Most drivers do not take the longest and most congested paths to work. They avoid these routes as much as possible. Basically, we are creatures of habit and our physical environment will dictate a lot of our actions and interactions.

As time goes by, you may begin to notice certain wear-and-tear on the road. Road signs begin to become less visual as paint fades away. Potholes appear and become deeper every day. Traffic becomes heavier as well and what once used to be a comfortable, convenient trip to work is becoming kind of annoying. Your physical environment is affecting you.

Finally, the city decides to utilize tax money and schedule road improvement projects. Of course, now begin the complaints about the change going on because we have to deal with detours and roadside workers. Funny, huh? You find yourself complaining about the improvements that you have been thinking of all along.

Once the improvements are done, maybe there now is a roundabout and not a stoplight. Maybe the road turns in different directions and there are road signs communicating different information. Speed limits are modified, bike lanes are installed, and new sidewalks are now available. This new physical "layout" will not allow you to go back to your old ways of driving.

Changing the physical layout of the office, warehouse, production floor, or wherever is important in breaking old ways of working. At some point, the culture will adapt and will almost forget what it used to look like. This chapter describes in detail why this is critical in your Lean journey as long as the move is to improve performance at some level.

Chapter 6 (Making Change with Lean) goes even deeper into improving your company and changing the culture. Concepts, such as cellular manufacturing, total productive maintenance, setup reduction, Kanban, visual communication, and in-line production, are discussed. Make no mistake, 5S is part of the overall Lean methodology and it sets the tone for the improvement that would be made through topics discussed in Chapter 6.

Chapter 7 (Keeping the Lean Fire Going) discusses the importance of training your people effectively and how certain approaches of training just do not work. I like to call it the *Lean lightbulb*. It is the psychological button that clicks on an internal switch that converts a person who is doubting continuous improvement to a full-on believer. This switch will come on for most people; however, it comes on at different times for each. But, some may never see the Lean lightbulb come on, and then some tough decisions need to be made by organizational leaders.

Once the fire is lit, the next challenge is keep that fire going strong. Keeping people motivated, balancing the day-to-day work, and continuous improvement can be difficult. Lean is a battlefield in a war that never ends. Of course, I choose to use the terms *battlefield* and *war* in the context of company goals, strategic planning, training, culture change, and winning in the competitive global economy. It outlines proven steps to keep your people motivated and making Lean a business model for long-term growth and prosperity. I also discuss how to make the transitions from leader to leader. Often, presidents, as an example, will be the initiators and vision setters for Lean. They will create strategic goals for Lean, invest in Lean training, initiate improvement projects, create a Lean culture and a Lean business model, and monitor results. At some point, these leaders leave the organization and their replacements decide to scrap the Lean journey altogether or take the company in a direction that does not truly embrace continuous improvement. There must be fundamental transitional steps for leadership to ensure the Lean journey does not fade. Positive changes within the Lean business model are fine and even encouraged, but a complete discard is dangerous.

So, you are now prepared to dive into the world of the human psyche and, more uniquely, this beast we call Lean. There may be points of pain while reading this, but this pain will be relieved and then eventually cured. For some of you, this book will provide the answers to the questions and issues as a leader you have been asking yourself. Any company preparing for or in the middle of continuous improvement will find the information you are about to read, informative, helpful, and enlightening. Equally rewarding will be that you find that the business advice outlined is directly applicable to your personal life as well. Enjoy.

Chris Ortiz
President, Kaizen Assembly

1

The Psychology of Change

Fear of change. We all have it. We all deal with it differently. Some of us accept change immediately, some take a little time, others never get there. I really want you as an organizational leader to look within yourself as a potential change agent for your company. Before you can get an understanding of the psychology behind Lean and waste, you have to deal with the way you think as a small/medium/large business owner or company leader. This chapter helps prepare your mind for Lean and leads you toward being a more forward Lean-thinking torch bearer.

PERCEPTION OF CHANGE

Change is never easy. Even in micro amounts, we as humans avoid change. Some are great at change, but you first need to change your perception of change and possibly what is it inside of you as a business owner or leader that is stopping change in your organization. Small business owners are great examples to use in this because (I am one, so I can relate) it is often the small business owners who strenuously avoid continuous improvement. The reason is that, as an entrepreneur and small business owner myself, we first have to admit that this little "baby" we created may not be all we think it is.

As a small business owner, you came up with the idea, started the business, grew it over the years, and feel that no one else really knows your business. I encounter this a lot in my travels.

Small business owners have a close connection to their companies that even executive leaders of corporations or other business entities do not. You can compare it to parenting. I have two wonderful sons. They are my babies; my wife and I are raising them in a certain way, and, of course, as parents we have a major stake in their success as humans. Even the most experienced nanny or babysitter, who may have been with us for years and knows the in and outs of our children, will never have that connection. Small business ownership is similar.

So in a small business, the first major obstacle in starting a Lean journey is the owner (owners). If you are a small business owner, you need to start thinking about continuous improvement, and maybe the way you have been operating will need to be changed in some form. As I have grown Kaizen Assembly over the years, I have become more and more open to changing how we operate. I struggled in the beginning as I hired people and gave them a say in how I ran the company. I noticed over time as I let go of that hold I had on my "baby" that things began to improve. It was a very fulfilling experience.

Small business owners often need to have their hands in everything and, in many cases, their decisions can hurt the company. Here is a good example.

A few years ago I met with a company that was in the business of assembling military equipment for the government. Most of its operation was true assembly as it really did not manufacture anything. Minor fabrication was performed once parts came into the plant, but mainly the staff were assemblers. It was a small facility and owned by someone we will call Paul.

From a psychology of change perspective, Paul appeared serious about turning his assembly operations into a Lean organization. When we had our initial meetings, I did not see any major obstacles in making internal improvements. Part of the early Lean assessment was to evaluate his company's supply chain. The company had a few suppliers that provided welded parts, powder-coated parts, and electrical components. Paul's welding supplier was someone he had been doing business with for many years and, during the early years of the business, the relationship made sense.

Volumes were low from the military equipment side and the small welding vendor could keep up. Due to long setup times and internal inefficiencies at the welding company, Paul was stuck with buying in large quantities and storing them in his small facility. It would take weeks before the lot of welded parts was consumed and it created obstacles for his assemblers. It was just messy. Sometimes the welded parts were incorrectly made and the welder would have to create new ones to replace the defective parts. Lead times were long because the welder could only be set up to work on one job at a time.

Paul also was required to pick up the parts because the welder did not deliver, and they both wanted to save money on delivery costs. So, Paul would have one of his assemblers leave the work area and drive to the vendor. This relationship was probably just fine in the beginning, and as Paul hoped, business picked up. The business elements around him began to change, but, oddly enough, nothing changed internally to adjust.

Paul's product offerings began to change, so he needed the flexibility of quick changeover to different product lines and faster deliveries from his vendors. Paul had a small facility; thus he did not have space for large lots of vendor parts.

As the Lean journey progressed within Paul's company, the lightbulb began to come on. He couldn't maintain the same supply-chain process. The welder was still requiring large lot orders, kept making mistakes, was consumed in long setups, and did not want to deliver. Soon Paul's buyer began to get a little vocal, as did the operations manager. "Are we doing Lean or not?" they would ask. It was time to meet with the welder and discuss a change in the agreement. Paul was quite nervous because he had been using this buyer for some time and did not want to burden him or hurt his feelings. Paul returned with no good news. Nothing would change. The welder was not willing to make any changes and placed the pressure back on Paul to deal with it. Paul agreed with the welder because he was scared of hurting the relationship.

Keep in mind that I am a major proponent of healthy, long-term business/personal relationships, but this vendor was hurting the long-term growth and profitability of the company. What happened? Paul rented storage space for the extra welded parts, rented a delivery truck to pick up the parts, and the welding supplier continued as usual. Not good. When the economic downturn of 2008/2009 occurred, Paul's company began to get cash strapped. There were more reasons than just a poor supply change; however, the moral of this story is that small business owners are guilty sometimes of keeping relationships with people or processes that are not helping them—either out of guilt or ego. If this welder was a long-term ally, he was doing Paul no service.

Now, as a company leader (not necessarily a business owner), you have to take the same approach to managing the company. You probably have less or no emotional attachments to the business, but you do have a stake in its success. First, admit that there is room to improve on a large scale and even in small increments. Maybe you need to question your own leadership ability and how you perceive change. I realize it may take time, but change your perception of change. Don't be scared; fear will get you nowhere. When you open your mind to change and continuous improvement, doors will open everywhere and there are endless possibilities of where you can take your organization.

VICTIMIZING

One of the elements of Lean and the perception of change is overcoming the concept of victimizing. Some of you who are experienced in Lean will know precisely to what I am referring. As you begin to change your perception of change for the better, a similar paradigm shift will have to happen with your employees. Victimizing is an early occurrence in Lean journeys and we see it in all work environments, especially in administrative functions. Victimizing, at least within Lean, is the sense people have that the company is reducing waste for no real reason; basically, making change just for the sake of it. It almost borders on a feeling of being personally attacked. A very common example is when Lean implementations include reducing the amount of tool or supplies to get better organized and to reduce cost. The Lean concept of 5S (sort, set in order, shine, standardize, sustain) or the visual workplace is a good example of this. I will discuss this in greater detail in later chapters, but the visual workplace is created to provide a clutter-free, aggressively organized work area where tools, for example, are on vertical shadow boards and easily accessible. The first step in getting there is sorting through the area and identifying what is needed to perform the job. As the 5S team is sorting tools and removing what is deemed unnecessary, I often hear from resistant workers, "What are YOU going to do with MY tools?"—You, Me, You Guys, Us, Them, etc. These types of comments are a clear sign of victimizing and why the bigger picture of Lean is not yet clear in the culture.

People become very attached to their space and, oddly enough, to things they don't own. In an office cubicle or in a production workstation, there is a sense of oneness with the "stuff" at their disposal. Often it is the only place at work that they feel they have control over. Office workers are much more attached to their surroundings than production workers or even maintenance technicians.

You can sense the anxiety in people when their work area is being changed and, more severely, when they feel that there is no real reason why (at least in their minds). Victimizing can be reduced in your organization once you become metric driven and instill a sense of purpose. The end of this chapter shows you how, but until then be ready for the finger-pointing to begin.

LETTING GO

Some call it delegation, I call it *letting go*. Letting go as a leader is difficult. There is a difference between delegation and letting go. Delegation is simply the allocation of the proper resources to get a job complete so that you can focus on other things. I delegate the printing and creating of training booklets to my project coordinator so that I can focus on running the business and supporting our clients. Creating training booklets is not below me as a business owner. It is an important aspect of the business and I could do it if needed. However, that is a responsibility I simply delegate to the coordinator, and she is good at it—so even better for the company.

Letting go is different. It is the releasing of the emotional attachments to the business that are hurting its ability to improve. More importantly, it's the letting go of your resistance to change. A good example of this is a client we have in Bellingham, WA. The company is a supplier of various military, diving, and other outdoor products and owned by a husband and wife. They started the business some 30 years ago, and it has grown into a market leader in its industry. The owners are nearing the end of their time with the business, and their involvement is decreasing every year. When initially contacted by the company, we were approached by the operations manager who was tasked with finding a Lean consultant to assist them. After our initial meeting with the operations manager, CEO, and CFO, we knew that these three people were running the day-to-day operation, but the owners were still involved. I wondered how the owners' involvement would affect the early stages of the Lean journey, hoping they would not be another "Paul."

I had a chance to meet with the owners over lunch one day. I could tell right away they were not going to be an obstacle. They were extremely open to change and trusted the people they had hired to take on the emotional side and the implementation side of change. Although with some minor bumps along the way, they are slowly turning into a Lean organization from top to bottom. Improvements are ongoing, results are showing, and the owners are just paving the way. They have let go of their emotional attachment to the business and thus make sound decisions.

This company also has a very large administrative function of sales, marketing, graphics, customer service, accounting, and purchasing. This openness to change is seen in every function of the company. Now, of course, as owners, they won't let the shop sink and get derailed; they have embraced change and are allowing the improvement to happen. Things could have been the opposite, with the owners making *all* the decisions and holding on to old ways of thinking, but they have refused to do this and, therefore, the sky has become the limit for their company. That is letting go.

LEAN IS A BATTLEFIELD

Our psychological analysis of Lean and change now heads into the trenches of your organization. When company leaders break their emotional hold on the business and steer it toward a path of continuous improvement, the people below them now need to do their own soul searching. Word on the street may be that the company is embarking on a Lean journey, but the employees will get their own taste during the actual projects. We now find a unique psychological battle.

There are two responses you will get from your employees when it comes to waste reduction. The first one is the fear of job loss. Improved productivity and reduced cycle times may be perceived as less work, and then less jobs. Unless the company is nearing complete bankruptcy, Lean is not intended to eliminate jobs. Second, you will find instant buy-in.

Resistance to change will come in a variety of forms, and we as consultants can see it at all levels. Frontline workers may or may not see Lean as "Leaning people out." The philosophy of Lean truly should be used to help grow the organization, secure and create jobs, become more competitive, and, most importantly, satisfy the needs of your customers. One of my favorite sayings to production workers when I explain why the company is going Lean is, "Your customers require it." It places them right in the face of those people and companies buying product or service products from them.

A key implementation technique in Lean is kaizen events. Kaizen means *continuous improvement* in Japanese; thus, a kaizen event is a continuous improvement event. It involves a scheduled timeline, such as five days, a selected team, a selected work area, and goals for improvement. During kaizen events, the teams are quite busy making changes. They might be rearranging the work area for better flow of product, implementing a visual workplace, creating procedures, developing new fixtures, designing special tool holders, building shadow boards, etc. Lots of multitasking, evaluations, hard work, and results happen during a kaizen event. The result could be that of improved productivity, reduction of floor space, and reduction in inventory. Kaizen teams are generally exhausted because the amount of work done is compressed into a short time frame. Afterward, kaizen teams will provide a tour of the new area and show off their results. And, rather than stepping back and really taking in what just got done, you hear, "Why did you pick blue bins instead of red ones?"

This is a good example of what you will have to deal with. People often find something "negative" or out of place to recognize, and not the good things. This is just part of the resistance to change even if it does not involve the person making the comment.

Change can be difficult. Even though positive transformation can result, changing paradigm, breaking old habits, and discarding established routines can be tough transitions for anyone—management included. We, as Lean practitioners, have been preaching for years about the importance of getting management's commitment to change. The necessity is as old as the Lean philosophy itself. Nevertheless, lack of commitment from management is the one, definable reason why Lean implementations fail. This is a well-known fact and, yet, many company leaders embark on a Lean journey with initial zeal, only to allow it to fall apart, like every other new idea or program.

Why does this occur? The answer is simple—maintaining the status quo requires very little effort. Business is good and the customers seem to be happy. Profits and cash flow are good, employee turnover is under control, manufacturing processes are running well, and operational conditions appear stable. It is when things are going well that companies often hesitate to implement something new, not wanting to risk what they feel is already a good thing.

If you think you may be perceiving your organization in this way, you must know that you have one, very large group of supporters out there, supporters who want you to maintain your established methods of operating and eliminate any thought of continuous improvement initiatives: your **competitors**. Your competitors want you to conduct business the same way as you are now doing. It offers them a fantastic opportunity to gather up more of your market share, so they are behind you 100%. Global competition is a fierce reality of the business world. When asked about why they chose to embark on a Lean journey, my clients commonly responded, "To stay competitive." The principles of Lean manufacturing can truly help your business continue to play competitively by operating within a framework of continuous improvement. Do you know why this is important? Your competitors do.

COST–QUALITY–DELIVERY

The focus of any business should be its customers and how the customers' needs can be met. Potential and existing customers all have certain expectations, and businesses must try to establish processes that satisfy and exceed those expectations if they wish to remain competitive. In today's global economy, people can go just about anywhere to buy products and services, which is why customer service has become such an important factor. The concept of Lean manufacturing is no longer new. It has made its way through a variety of different industries and has been proved successful. Even so, many companies have still not fully embraced the Lean philosophy or applied what they have learned about it to their daily operations. And, yet, these same businesses may wonder why their company performance is in jeopardy as compared to their competitors. Athletic teams and athletes operate under the principle of continuous improvement, constantly seeking to find an advantage or competitive edge that will make them more successful as a team. Operating a business is no different. Basically, there are three main business decision drivers in today's world that can make or break a deal. When customers come to you seeking your particular product or service, they will base their decision on cost, quality, and delivery. Cost, quality, and delivery are the metrics used by a potential client when deciding if your product or service is worth purchasing. It is important to find an efficient balance between these drivers, focusing on all three equally, so as not to give one all the attention and allow the other two to fall apart. However, optimal balance of focus can be difficult, and each company defines it differently. The only way to ensure a focused balance is to employ the methodology of continuous improvement, which is the fundamental principle of Lean manufacturing.

Lean manufacturing principles, when correctly applied to your environment, can be a powerful tool to determine optimal cost, quality, and delivery for your products or services. Although there may be a few businesses that always look for the cheapest method, the ideal approach is to find balance, which takes focus and dedication. In the following sections I break down each of the drivers and note how they can affect each other within a manufacturing environment.

Cost

Typically, companies will handle cost in one of two ways: cost cutting or managing cost. Those that favor cost cutting employ downsizing, firing, improvising, abuse of suppliers, and cutting corners, all of which are clear cost-cutting actions. Some companies even believe that Lean manufacturing is cost cutting, which is simply not true. Rather, Lean manufacturing is all about *managing* costs, which, in the long run, makes much more business sense.

Focusing only on cost, the quality and potential delivery of a product or service will suffer. In an attempt to lower the cost of labor, many manufacturers try to limit the number of hourly workers and run "thin" assembly lines. Both of these measures constitute "cutting costs," but they are usually futile efforts. What typically happens in these cases is that scheduled workers are required to pick up the slack and perform work more quickly in order to meet critical deadlines. Essentially, the process is working beyond capacity, and this is dangerous because critical assembly and quality steps may be skipped in order to meet a delivery date. What typically occurs is that the delivery date cannot be met due to a shortage of personnel, and the product is either shipped late or with substandard quality, upsetting the customer and tarnishing the company's image.

As a Lean practitioner and trainer, I advocate making process improvements without a lot of cost; however, there are certain necessary items that must be on hand to ensure an effective process. Assembly lines and other manufacturing processes all require tools, workbenches, conveyors, lighting, part presentation, documentation, shelving, bins, tool holders, equipment, fixtures, and jigs, and these items must be working properly to allow the product to be built smoothly and efficiently. At some point when preventative maintenance efforts are exhausted, companies must replace items. Line workers cannot perform efficiently using unreliable or broken tools or equipment, which many are forced to do in order to save cost.

The costs associated with improvements in training and mentoring are sometimes considered too expensive. Not smart thinking. In contrast, I recall a situation in which I was negotiating with a company in Burlington, WA, in 2005. We were discussing our potential agreement when the owner made it clear that the cost of initial training was something he couldn't afford NOT to do. As far as he was concerned, training was an absolute must in his ever-changing market. Needless to say, this company is prospering in an industry where those who are not embracing change are struggling. The owner is beating the competition and meeting the ever-changing needs of his customers. Manage your costs; don't cut them.

Quality

Without quality, you have nothing. Forced to choose just one driver to keep at an optimal state, I would choose quality, hands down. Now, by no means should cost or delivery be allowed to fall apart because of quality; but, in my personal opinion, customers are more loyal to quality than to any other driver.

Like all of you reading this book, we are consumers seeking products and services. With the occasional exception, of course, I truly believe that people are loyal to quality above all else. If the quality of the product or service meets or exceeds expectation, we will sacrifice a bit on the cost or delivery. After all, it may be worth a little delay or additional cost knowing that we can rely on that particular business to provide good quality products and services. I do realize that there are many consumers who simply look for the rock-bottom price, yet I still find that most people value quality the most. Let me give you a personal example.

I own a car, as do most other people in America. Owning a car carries a variety of associated costs, such as gasoline and insurance, for instance. And, if we want our vehicle to last, we must spend money for preventative maintenance. As a consumer and automobile owner, I can take my car anywhere for repairs. There are countless options, and each automotive repair company offers some level of cost, quality, and delivery. That being stated, sometimes we opt to have our buddy repair our car rather than take it to a stranger, although most of us don't have that luxury.

There is a quick-lube shop down the street from my house, probably five minutes away, at most. This shop offers a multitude of services and products, and at a very inexpensive price (cost). They are very fast and can have my automobile done in a timely manner (delivery). The problem I have with this particular shop, which is not representative of other quick-lube shops, is that their quality is not to my satisfaction. They tend to skip steps, forget services, and make errors in the invoice. They also make a practice of hiring high-school kids with limited knowledge of automotive repair. My point is that the quality is not good, in my opinion. Although they appear to be balancing the other two drivers quite well, quality suffers and I am loyal to quality. So, what is my alternative?

I choose to take my car to a repair shop farther away from my home. This shop tends to take a longer time to get the work done (delivery). Therefore, many times I have to wait in the customer lounge reading magazines. Their prices are higher than most shops in town (cost), but their quality is outstanding. They have fantastic customer service, and I never have an issue in regard to their quality of work or documentation. Basically, I am willing to put up with a longer delivery time and higher cost because I know the quality of their services is worthwhile. However, if cost and delivery got farther out of balance, I might be tempted to seek out another vendor. That is a real possibility.

Now that you have a good image of quality, I would like to discuss what happens when too much effort is spent on quality of product or service. In my repair shop analogy, I was happy with the balance of the three drivers, although this was a personal preference and may not reflect how other consumers might feel. But, whenever there is excess focus on quality, cost and delivery can suffer.

Manufacturers that focus most of their attention on quality may, in fact, not trust their ability to manufacture a quality product. Perhaps there has been a history of poor quality and they are afraid of losing customers. Companies that spend a lot of time on quality have a greater chance of increasing operating costs. Assigning more people to focus on quality is expensive and, at some point, it becomes redundant. Inspections and testing are not considered value-added activities unless the customer is willing to absorb the cost. Some manufacturers have government contracts that demand strict adherence to specific criteria, but they are willing to pay extra for that. In this particular case, it may be cost effective to perform repetitive inspections and testing, but most manufacturers do not fall into this category. Delivery also is negatively affected by an over emphasis on quality checks, with extensive lead times due to the extra time needed for checking and rechecking product.

Delivery

Are you volume driven? "Build, build, and build" seems to be the mantra of every manufacturing company. In my 10 years as a Lean practitioner, I have not yet come across a manufacturer that was not output driven. This is not meant to imply that making deadlines and delivering the product on time are bad things. It is the ridiculous notion of speed that seems to be the cause for much unnecessary chaos. Companies that focus on delivery exclusively will very quickly see an increase in cost and a decrease in quality, especially in uncontrolled working environments. Companies that have this volume-driven mind set cannot see any farther than the next hour; there is no long-term vision for the day, the week, or the month, and crisis management is the name of this game. The first indicator of an inefficient line layout is a production supervisor or manager who "throws people at the process" to make daily production requirements. Adding people to the process, without reason, costs money and will not add any inherent value to the flow of the product. The cost of labor simply increases, as does the risk of potential quality problems, such as scrap, rework, and future customer complaints. It is difficult for many managers and line operators to break this habit, even after Lean principles have been applied, but it is a mentality that should be reversed for many reasons. The added cost is not usually passed on to that first customer, but, eventually, it may be added into overall business costs and that will affect future customers. It is a cost that must be initially absorbed by the manufacturer; therefore, it may be pulled from other areas of the company, causing a shortage of funds there. Business 101 is to establish your organization to accommodate the original price agreed upon with the client while being able to fund operating costs. Therefore, you are just shooting yourself in the foot by creating added costs outside of your business structure, and if you compound these costs for each line in the plant for one year, it begins to add up. Additionally, you will need to prepare for the external costs that could arise through customer complaints, warranty work, and service calls.

Simply put, there is no gain to an organization if it simply creates problems and adds cost by driving processes and people to meet unrealistic delivery dates. Manufacturing processes should be designed to meet delivery dates effortlessly, and should be monitored continuously to meet that objective.

Balance is the key with cost, delivery, and quality. Achieving optimal balance in these drivers will be an ongoing challenge. Lean manufacturing provides a variety of tools and principles that can be applied to a business to help achieve optimal cost, quality, and delivery. Your first step in getting there is to recognize that the customer can go anywhere to do business, and, in fact, may have already done so. Customers are the reason we exist; so remember that they have a choice, and their choice will always be based on cost, quality, and delivery.

CHANGING YOUR BUSINESS STRATEGY FOR LEAN

Commitment to change is essentially a philosophy when it comes right down to it. However, it is also the key to successful Lean implementation. Change is not an easy thing for some individuals. In truth, for some people, change is extremely difficult because it rocks the boat and makes things that were familiar suddenly become unfamiliar. Getting people to embrace change as a positive measure is very difficult as well. There is no perfect template to follow when adapting to change or teaching others to adapt to change. The challenge lies in getting everyone to see change as valuable and necessary, and, ultimately, the best insurance for success. Effectively, change is not in our words—it is our actions.

To avoid Lean implementation failures on the shop floor, leaders must first establish the foundation for success by developing goals and metrics to improve **cost, quality,** and **delivery.** Improvements in these areas will have a profound impact on the company's financial strength as well as overall growth. Actions should always follow words. Don't simply state, "We are doing Lean," demonstrate it by taking action. This important concept is what I call the *strategic purpose.*

The strategic purpose is an effective way for management to demonstrate their commitment to the Lean program and to the Lean philosophy of continuous change and continuous improvement, beyond just inspirational speeches and company declarations. The strategic purpose serves as a guideline to implement Lean processes and sustain positive change. Some individuals refer to the strategic purpose as a Lean strategy.

Each company needs to establish a list of critical shop-floor metrics that can be measured and quantified in order to help the business overall. On the production floor, these metrics are often called key performance indicators (KPIs). Creating these metrics is just one part of creating the strategic purpose.

- Productivity
- Quality
- Inventory/work-in-process (WIP)
- Floor space use
- Throughput time

The six middle metrics are critical in measuring your success with Lean, and it is these metrics to which most people can relate. It amazes me the number organizations, including some pretty well-known companies, that are making products and they do not measure anything. I want to take a moment here and describe what I call the *psychology of metrics*.

In the early phases of Lean, you must begin to instill a sense of measurement in the minds of all your employees. In manufacturing, auto body, food processing, etc., you have to develop metric-driven people. Once people have an understanding and an appreciation for metrics, the improvements made will have more meaning for an organization as a whole.

Productivity

Productivity is often confused with speed. Work faster! Faster compared to what? Speed will get you nowhere fast. When managers instill a backward philosophy of speed, they are only creating an environment for potential quality and safety issues; basically focusing on delivery. You have to begin to teach and train your people the concept of pace. Pace wins the race in the long term. Even with brief and random episodes of speed, pace is the best thinking. When I witness speed-driven workers, it tells me two things: inefficient process and poor leadership. Now, on the other hand, there are organizations that do not establish any targets or metrics for people and, thus, they do what they want. This also is a sign of an inefficient process and poor leadership. There is a balance between running the company and creating an environment where people have input. I know, it's a tough balance. Metrics, like productivity, are important for your Lean journey because they can be used to measure your success and people can recognize the improvement. For instance, if you place a goal of 90% productivity to a standard and the line, on average, is operating at 72%, you now have a goal to reach. As a front-line worker, there is now a target in front of you everyday. Without a goal, there is no need to improve or try harder. After Lean improvements, we can see how we affected productivity and now show this improvement to the people. Once they see the connection between improvement and better productivity, then you are instilling that sense of accomplishment in them. Then, keep it going because they are improving what can be measured in a variety of ways. Productivity is improved when products are manufactured with less effort, less manpower, less equipment, and less utilities (overhead). Because Lean is about managing costs—**not** cutting costs, manufacturers need a smart approach when attempting to achieve minimum effort. Supervisors and managers always seem to be concerned about speed, but productivity is not about speed; it is about pace. Over the years, manufacturing professionals have had a misconception about productivity. Working hard and fast to excess is not conducive to good quality and safety. Human beings can only sustain 100% speed to a point without negatively affecting quality or seriously negating safety factors. Lean is about working smart, at a pace that is sustainable, while safely producing the required number of good-quality products in a given time period. There is a methodical approach to designing a process that creates a smart pace, allowing managers and engineers to calculate accurate manpower requirements.

Productivity is directly related to cost. Therefore, it should be the most important shop-floor metric to have in your strategic purpose. Focusing on the efficiency of the production operators is critical because they are considered value added; they build product, which, in turn, pays the company's bills. With a more efficient process and less personnel, cost will be dramatically improved. Less personnel equates to less labor, less tools, less workstations, less documentation, and less material, for example. Fewer people handling product equates to improved quality. These are key points that contradict the old philosophy of simply throwing more bodies onto a line in order to meet deadlines. Unnecessary people working on the product, in an uncontrolled and poorly designed process, can increase the chance of quality errors—not to mention safety issues. Working in a Lean environment means *less*. Less processes, less people, fewer transactions and systems, fewer places the product must travel before landing in the customer's hands. Delivery is improved substantially. The important thing to remember is to manage costs and redeploy personnel as needed when areas become more proficient due to improved processes. Don't just cut people. That is not the Lean philosophy. Proper manpower allocation is the best approach.

Quality

Although quality is already one of the three main drivers, I recommend it also be one of the key shop-floor metrics in the strategic purpose. However, before adding this to the list of metrics, a decision must be made: how will quality be measured? Surprisingly, I have witnessed many companies, large and small, that do not measure quality; a big mistake, in my opinion. Quality should be measured both internally and externally. Some companies use customer complaints as their external measure. They simply keep track of the number of monthly complaints and the cost of problem resolution, i.e., warranty costs, cost of service calls, etc. While those costs are important, internal costs also should be tracked, i.e., rework costs, scrap costs, parts per million (ppm), number of rejections, etc. Quality can be measured in a variety of ways, and it truly depends on your products and processes. Regardless of which method you choose, a measurement of some kind must be in place.

It will be difficult to improve quality if the production process has a large degree of design variability, a lack of cross-training among line operators, unreliable equipment, outdated documentation, or poor flow. Quality is improved when there is a reduction in the number of defects, rework, scrap, and external complaints. With this error reduction, less time and money will be spent reacting to problems after they occur. Delivery will be improved, due to less downtime created by stoppages on the line. A controlled manufacturing environment promotes good quality, with more predictable lead times—factors that will definitely please the customer.

Inventory/WIP

A lot of money is tied up in parts and material, which is why they should be part of your metrics and providing a method for monitoring and improvement. However, reduction of inventory takes time. Pairing down inventory to a satisfactory level is not a process that can be done overnight. This particular metric also includes work-in-process (WIP): parts, subassemblies, partially completed units, and finished goods. The ideal state is when completed units have been loaded onto a delivery truck and are not just sitting around in some physical or logical location known as finished goods inventory. Unless your plant is used as a distribution center, fully built products should leave your facility relatively quickly.

Many manufacturing facilities literally use their production lines as stockrooms, with large quantities of parts and material lying around in the assembly area. Some managers view this as a positive method of reducing indirect labor costs because fewer workers are needed to bring material to the work area. Although this might make sense in the short term, a much higher cost is incurred by ordering larger quantities of parts and then storing them. By allocating a smaller, controlled number of parts and material into the process, a company can afford to have personnel to manage it. This cost will truly outweigh the cost of excessive inventory.

Excessive inventory and WIP also create other problems. First, there are storage and handling costs. Large lots of parts or WIP can create quality problems due to over-stacking, forklift damage, and hidden product defects that are undetectable in large lots. A huge infrastructure is required to control and monitor excessive parts and WIP. Inventory takes up valuable space that could be used for more profitability, such as the production of a new product. While I may be painting an ideal picture, it is important that each organization find that healthy balance between part quantity and the cost of indirect labor. Using inventory as a metric in the strategic purpose is a critical part of achieving good cost, quality, and delivery. Cost is significantly reduced if a company does not have money tied up in excessive inventory or WIP, which is a burden on business operations. Quality is significantly improved, damage is less likely to occur, and initial defects can be better visualized in the smaller lots. Delivery is also improved because time is not spent moving large lots from one location to another, or managing finished goods that already could have been shipped.

Floor Space Use

Floor space comes at a premium, and employees need to start looking at the poor use of floor space as hurting the company's ability to grow; growth that creates job security and self-enhancement. Renting, leasing, or buying a manufacturing building is one of the highest costs of overhead. Utilizing floor space efficiently is critical to the success of any Lean journey, especially in regard to cost, quality, and delivery. The production floor serves one major purpose: to build products. Although the factory is used for other items, such as holding inventory, shipping, receiving, maintenance, etc., the production floor should be effectively utilized for value-added work: building products. Non-value-added work on the floor means less profit for the company, especially considering the high cost of owning, renting, or leasing the building. Over time, items such as workbenches, garbage cans, chairs, machines, tools, tables, carts, parts, and pallets tend to accumulate and valuable production space disappears. Including floor space in the strategic purpose is critical. As metrics are monitored and improvements occur, valuable space will become free, allowing manufacturers to add more production lines, produce more products, and become more profitable.

If vacant space exists, without clear, visual objectives for use, the space will get filled. That is just human nature. Unnecessary items will begin to pile up and consume production space, either on the floor or in the work areas.

Cost is significantly improved when floor space is better utilized. Adding production lines for new products brings in additional revenue. Or, if through metric measurements, a company determines they need less floor space, they can move to a smaller location, saving money on their rent or lease agreement. Quality is significantly improved when less space is used to store unnecessary items, which could become damaged or broken. Also, less clutter in the area promotes better visibility throughout the factory. Delivery is significantly improved through the reduction of unnecessarily long assembly lines and other manufacturing processes that could make more efficient use of space. Products and parts spend less time traveling around the factory, which means shorter assembly lines, shorter processes, and, ultimately, shorter lead times for customers. Does it make sense to spend thousands of dollars on monthly lease payments for a factory floor that is only 50% utilized? Floor space is expensive, use it wisely.

Throughput Time

Sometimes used in conjunction with travel distance, throughput time is the time it takes the product to flow down the production process. Obviously, throughput time has a direct impact on delivery. The more time a product takes to go through the main process, the longer a customer must wait. Of course, there are a multitude of variables that can extend product lead time; therefore, it is wise to simplify the metric by monitoring the throughput time. As an alternative, travel distance can be measured instead of throughput time. It's up to you.

Longer production lines require more workstations, workers, tools, workbenches, conveyors, supplies, parts, and material, which results in additional costs and WIP as well as extended lead times. I have witnessed some impressive improvements to throughput time—travel distance reduced from 300 feet to just 35 feet and from 550 feet to only 100 feet. A physical reduction in distance equates to less throughput time, allowing an organization to promise competitive, yet reasonable, delivery dates. Also, from a visibility standpoint, it allows production supervisors to see everything taking place in their area from a single vantage point. The less time the product spends in the building, the better. By improving throughput time, a company benefits in terms of cost, quality, and delivery.

Improving key shop-floor metrics will have a profound impact on the overall financial success and long-term growth of the organization, and this is something your people need to understand. The six middle metrics discussed earlier are particularly important because they are directly related to the work being performed on the production floor, which affects profit and, ultimately, employee paychecks. Production workers need to work in an efficient environment in order to be successful contributors to optimal cost, quality, and delivery. Basically, they can have a positive impact on the business through their efforts. Pretty cool, right?

Another way to change thinking is to place a financial measurement to floor space. My favorite is profit per square foot. Because floor space predominantly should be used to create revenue, it essentially is creating profit. The less space used for value-added work, the less profit per square foot the company can acquire. As Lean improvements open up floor space, rather than simply measuring in terms of square feet, use profit per square feet. It becomes opportunity profit, potential space for creating revenue. The company will look at floor space completely differently now.

The shop floor is only one part of the overall Lean strategy. Departments, such as finance, accounting, customer service, purchasing, engineering, shipping, receiving, and many others, also contribute to achieving balance between cost, quality, and delivery. This approach to Lean enterprise, while a bit unconventional, makes good sense. With a strategic purpose in place and metrics established for the production floor, employees have a better feel for why they are in business. Although each department should eventually develop its own internal metrics for improvement, when it comes to developing an all-encompassing strategic purpose, the shop floor is the key.

PSYCHOLOGY OF METRICS

Developing your people into a metric-driven culture will help channel their focus and allow them to have a different perspective on their role in the company. I am not implying that they will become a metric-driven culture overnight; but as you chop away at it, they will eventually clue in to the reason why Lean is in their life.

A business is in place to create profit, pure and simple; it is not a charity case. Being metric driven does not mean that as a Lean leader you lose sight of balancing the sometimes nonmeasureable side of a company. Ethical work environment, community image, work–life balance, environmental stewardship, and providing a place to grow and prosper as individuals are all part of operating a company in society. Recognizing and creating these elements of a company is part of being a Lean leader. However, it is my professional opinion and experience that metric-driven companies are more eager for Lean and have targets for their improvement efforts.

When Lean tools are applied to reduce waste and improve performance, and your people can see how their efforts improved something measureable, you instill that sense of goal thinking and overall team contribution. I want your employees to leave work every day thinking that they contributed to improving the business through their efforts. And, they will have measurable results to prove it.

2

Leading the Lean Journey

Taking the role of a Lean leader is a big step. Now that your perception of change is more forward thinking and you are aware of the resistance you will have to deal with, I want to explain in this chapter the finer detail of how to lead that Lean journey.

WHAT HAVE WE DONE FOR OUR CUSTOMERS TODAY?

Knowing your customers is the first step in being a Lean leader, and even more important is instilling a sense of customer expectation into your employees. Many organizations today have truly forgotten who the customer is and why they are in business. You are in the business of providing a service or product that satisfies the customers' balance of cost, quality, and delivery. Your operation must have processes in place that satisfies exactly these objectives.

Creating an environment of "customer need" is healthy and critical to being competitive. Some industries are very close to their customers. If you are an auto repair or auto collision repair shop, you are very close to your customers. You have constant interaction with them at the estimating and administrative point. Technicians are working diligently to get the repair done in an efficient manner so the customer can use the car again. Most service industries are close as well, which includes what Kaizen Assembly does. We work side by side right in the heart of the operation in driving change.

Manufacturers have some distance from their customers and I mean those on the production floor who are building product. This includes the maintenance staff, R/D, engineers, and other directly supporting production needs. Sales staff, customer services, and executive leadership generally have a closer connection. What you must do is help change the thinking of those who are not as close to the customer in which all of their actions affect the end customer in some way. Are they doing what the customer wants them to do? Buying excessive inventory is doing nothing for the customer. Neither is walking around looking for tools, moving product throughout the facility, setting up a machine, building poor-quality products, or taking longer to fix a machine. In Lean, you want them to be aware of the waste they are committing and then convince them to find ways to reduce or even eliminate it. Not improving the process also does *nothing* for the customer. I don't expect a second-by-second analysis of every single action, but rather a general sense of what adds customer value. Be a customer-driven organization.

DANGERS OF NOT GOING LEAN IN A DOWN ECONOMY

Another challenge, but also a learning opportunity as a Lean leader, is knowing when to improve a process and when not to. This sounds kind of self-defeating, but there is a company to run and you have to learn to balance the day-to-day needs and continuous improvement. There are four thoughts here and I want to hit on all four:

1. Business is good, why improve?
2. Business is good, we must improve.
3. Business is poor, why improve?
4. Business is poor, we must improve.

Business Is Good, Why Improve?

If you think this way, then your reasoning may be borne out of fear. I have come across a few leaders in my time who think this way and are happy with the status quo. Revenue is good, costs are down, margins are healthy, and people are happy ... why improve? I can see the rationale behind this, but it is not forward thinking. What a great time to invest in Lean and make it even better. Often when a leader has this mentality, the business is busy, work is steady, and there is fear of disrupting it. Why do anything to the car if dials indicate it's running smoothly? Well, this is a great time for Lean: look at the potential growth coming, evaluate your processes, and find ways to improve even more. Wouldn't more profits and lower cost be better? Now, on the other hand, a lot of firms begin reinvesting in the company at this stage because they have the money and resources to do so. Time may be an issue because employees may be swamped keeping up with the work and, thus, adding continuous improvement projects may be overload. As long as companies grow and increase profits, they often don't see waste. Sometimes organizational leaders are aware of the waste but do nothing. Left unchanged, the process just inherits the waste as "part of the process," and this added cost is now looked at as general operating cost. Good times could be blinders to problems.

Business Is Good, We Must Improve

This is the right mind set and everything is in your corner. This falls under the model of using Lean to achieve growth, create a stable workforce, and improve things even more. This is Lean thinking and the proper leadership quality. Taking time to stop and look at the company during healthy times is smart. Like the scenario above, healthy revenue often means a busy company. As a leader under this theme, you may be asking the same types of questions. Why should I add yet another layer of work with continuous improvement projects? Part of your Lean journey is evaluating your processes, collecting data, and capturing what we call the current state. Successful evaluations of the company happen when there is actual work to analyze. If your efforts are centered on a production process, well, there must be product being built. If office improvements are desired, office workers need a healthy stream of work orders, purchase orders, customer calls, etc. to watch and analyze. So, starting Lean when you are quite busy is great for that purpose.

Kaizen Assembly has a client that fell within this category. The organization also was experiencing some healthy growth and began to see that its current processes from marketing, customer service, production, and warehouse would not be able to operate in its current state. The pain of steady work added to continued growth was reaching a breaking point. I did a full evaluation of the company's internal value stream, which was a macro look from receiving customer orders to shipping the product. Major overhauls were needed if they were going to successfully capture and process the added orders that would be coming in during the next couple of years. Although they were quite busy, they went forward with a complete Lean transformation of the business, including an entire layout change of the office, production, and warehouse. It was crazy at times, but they achieved remarkable results.

Business Is Poor, Why Improve?

This is by far the most common scenario when it comes to avoiding change and continuous improvement. I have held various high-level positions in companies and now as a business owner, I understand the hesitation here. Money may be tight, with very little to invest. The most recent economic downturn revealed a lot about the power of Lean and continuous improvement. During this time, we did not lose one customer. We had a few contract reductions, but we stayed quite busy. The level of our workload is contingent on our customer's revenue stream. Companies implementing Lean principles often (not always, but often) can blaze through a slow economy because they have embraced Lean. The company does not buy excessive amounts of inventory and tie up needed money. Its workplaces are extremely organized, and it challenges the need to buy stuff unless it is absolutely necessary. The organization believes in a highly cross-trained workforce where people have acquired multiple skills. Machines and equipment have been maintained through comprehensive total productive maintenance programs, which reduce cost, spare parts inventory, and downtime. The production processes are streamlined with less waste, reduced travel distance, standard work, and less work in process. This all equates to higher productivity and quality. I could go on and on, but one of the great attributes of Lean is better cash flow and shorter inputs to outputs. Simply put, faster return on the investment in labor and material. When the economy drops off, yes, they may need to make some cuts, but maybe not as bad as if they were not Lean.

One of our longest-running customers witnessed this scenario. After four years of what I would consider heavy Lean implementations, the recession hit. I spoke with the vice president of operations and he stated that some cuts did happen, but they were not nearly as severe as others in the area experienced. He attributed this to what he called "a small bump in the road," rather than a major downsize due to Lean and the continuous improvement over the past four years. He also stated that when it did slow down, they accelerated their Lean projects. Long-term Lean journeys are, at their core, strategic and critical to surviving a recession or a slower economy. When times are slow, it is the best time to sit back and take a look at the big picture. What can you do to improve the business now so that you are prepared when it picks up again?

Business Is Poor, We Must Improve

Reality must be kicking in here if you are in this scenario. Now, this may mean you have reached a catastrophic stage of the business, or it may be just slow, and you now have a great opportunity to evaluate the needs of the business. Time may be available to do so and the change can be small, but it is yet another good time to begin Lean. As mentioned before, I am a firm believer in preparing your operation when business picks up again because in most cases it does. As of this writing (April 2011), the housing market is still on shaky ground and unemployment is still hovering around 9%. However, there are several positive economic indicators with manufacturing picking up; the Dow Jones Industrial Average is near 13,000 and consumption has increased. Most companies are not hiring due to fear. I can understand this as an employer. Adding payroll is about evaluating risk and no one wants to hire someone and then turn around and let them go in eight months. Turnover is expensive. We also juggle risk when deciding to invest or innovate. The innovation I am suggesting here is in regard to Lean. Let's be honest, a Lean journey requires a financial investment in such things as consulting, training, supplies, contractors, employee involvement, etc.

I can think of two companies we support that are aggressively implementing Lean during a slow time for them with this mindset. "Reduce cost, open up floor space, and cross-train everyone." The cost reduction is not in resources, but inefficient processes, excessive inventory, and poor suppliers. When business starts an upward trend for them, they will need to hire again and their current processes will not enable them to handle another dip in the economy. They have come to the realization that they need to be better prepared next time. Cuts to the workforce will not be needed or as drastic if they hit the waste reduction path.

Beginning a Lean journey and embarking on continuous improvement is a bold move. Trying to find the right timing for it is nearly impossible. You can juggle all four of the above scenarios but, truly, there is no right time or wrong time to start. Just let it begin.

CHANGING INTO A LEAN LEADER

Leadership can change you either positively or negatively. It can make you either a demon or a saint, and it will make or break Lean in an organization.

A colleague of mine used to be a plant manager and he earned a reputation for excellent treatment of the people who worked for him. After several years as their manager, he was promoted to a vice-presidential position with the same company. Six months into his new position, we got together for lunch to discuss my plans to write this book. My friend was a very different person from the man I had remembered. Corporate life had taken control of his identity and he seemed genuinely disappointed with himself.

I remember clearly one of the things he said to me that day. "When you accept an executive position in corporate America, you have to leave all your ethical reasoning behind you." He explained that while working at the plant level, he still had some control and was able to create a pleasant work environment for his staff. Now, as an executive, he had to play the role of a greedy, selfish, and ruthless businessman. Although he did realize that he could simply walk away from the company, this wasn't easy for him to do because the money and career opportunities were very beneficial. Therefore, he found himself in a continuous struggle, trying to maintain his former value system in a position that was in total opposition to his personal beliefs.

I realize his situation is not representative of all companies, but it got me thinking about guiding employees through a Lean journey. Can a company's past behavior toward its people affect the success of Lean? Are employees forced to work excessive overtime and placed into positions that make them unsuccessful and, then, over time create a negative culture. His company was beginning its Lean journey and he was finding conflicts with the way the leaders operated the organization and how they were being taught Lean by Lean practitioners like me. Lean journeys require investment, time, commitment, patience, and acceptance of mistakes. Managers controlling the day-to-day operations of the company were far from this approach to business, and he eventually found himself looking for another job. It was a personal choice.

My personal experiences over the past 13 years in the Lean field have taught me a lot of valuable things, especially about how to treat people. The companies I have assisted quickly realized that a new approach to leadership was going to be needed to ensure success in their Lean endeavors.

I was, by no means, a perfect employee the years leading up to starting Kaizen Assembly, and, in fact, was a bit resistant to Lean as well. However, I always maintained the belief that my resistance was normal, and I appreciated my great Lean leaders. How we treat people in our Lean journeys is the cornerstone of Lean leadership.

I took all that I learned from my experiences and use it now to lead companies in a manner that seems fair and just, and hope this approach trickles down through their own organization. Organizations embarking on Lean need effective leaders who understand the importance of employee contributions, and how much their efforts and attitudes affect the success or failure of a company. Certain corporate leaders need to realize that while aggressive practices may achieve short-term financial success, they also place the company on the path of a precarious future.

My colleague's perspective on leadership was altered dramatically after just a few short months in an executive position. Although he realized the negative personal changes that were occurring, he simply had no choice but to acquiesce—to conform. But as Lean was brought into the picture, he knew it was time to leave. Many executive leaders are breeding a middle management culture that is willing to sacrifice the rights of their employees. This is an opinion developed from my own observations. Again, not all leaders, but a good handful of them have led their organizations with too much negative reinforcement. Although profits are necessary, to help deal with the culture change in a Lean organization, this mind set can become an obstacle to success. Once the management level of a company has been indoctrinated, younger leaders are then trained to be loyal to the company at all costs, relinquishing their personal lives. The company becomes their lifeblood, and their identity becomes defined by a prestigious title and by their "company loyalty and dedication," which translates into how much of their personal lives are expended at work.

Lean leaders are only human beings. Therefore, they typically conduct themselves in a manner that reflects their authentic personality. If individuals are generally grumpy and negative toward change, their management techniques will demonstrate that, and they will affect the morale of others through their body language as well as their words. Speaking negatively about Lean around your people will not generate a following. This also is the case with your middle managers and supervisors. You are going to find a mix of resistance and buy-in; and if you have middle managers bad mouthing Lean, you need to address the issue. Their negativity will only wear off on their direct reports.

Individuals who are happy and positive tend to lead in the same manner. Lean leaders who do not let negativity influence their actions will create a following of positive thinkers.

In a way, today's leaders have lost a sense of reality. After all, treating people ethically is not like performing open heart surgery. It does not require week-long seminars and workshops to learn how to be nice. I enjoy leading people because of its simplicity. While leading a company is a monumental task, leading a group of employees is relatively easy. What follows is my concept of how to make Lean leadership easy and pleasant for yourself as well as those you lead.

Acknowledge and Involve Your Staff

When members of your staff do a good job, notice it and praise them for a job well done. Many managers don't praise employees because they feel they are paid to do their job well, and praise is unnecessary. That is a copout. It doesn't require much effort to say: "Good job." That is all that is required. Don't let an opportunity go by to praise a piece of good work.

Go to your employees for advice and engage them in problem-solving issues. Using the talent that is available to you is a key ingredient in building a positive and helpful team. One person may have the solution, or perhaps the department can resolve the problem together. In either case, involving your workers promotes trust as well as professional interaction.

Provide an Environment in Which People Can Be Successful

Train your employees adequately and give them all the tools they need to be successful. Explain all job responsibilities clearly, and encourage questions and feedback. New employees need your attention and deserve your support and encouragement while they are learning. Don't leave them hanging, or looking for answers, because it will indicate that you are not a manager on whom they can rely. Be sure to spend sufficient time getting to know individual contributors, so that you are able to assess their skills accurately and make sure they are assigned appropriate responsibilities and are challenged and inspired by their work.

Do Not Humiliate Anyone Who Works for You

If you are annoyed with someone on your team, or someone has done something wrong, keep your cool and bring it up when you are alone with that person. Embarrassing your people in front of others does not show that you are a good leader, but exposes you publicly as a tyrant. Always wait until you have an opportunity to discuss the sensitive issue with the employee in privacy.

Create an Environment Where Mistakes Are Okay

Mistakes are learning experiences and should be embraced as such by leaders of an organization. Typically, mistakes are made when an employee is in a learning curve. Beware of those employees who never make mistakes because this is usually an indication that they are not stretching themselves or reaching out for new opportunities. Treating mistakes as part of growth rather than something to be ashamed of allows your employees to feel comfortable taking risks on your behalf. "Take one for the team."

Do Not Hide behind Your Position

Be genuinely friendly with your people. Don't hide in your office during the day, but make yourself visible and available. Be a visible leader. Stop by the cubicles or offices informally just to say hello, or find out how everything is going for each of your people. You would be surprised what can be gained from this type of interaction. Standing as a team allows you to support and encourage each other when times are tough.

Be Approachable

Maintain an open-door policy. Allow your staff to come to you whenever they need to talk about sensitive issues, difficulties outside of work, or even just simple small talk. This is one of the most valuable leadership tools I learned from my manager at Fibrex. Whenever I walked into Ronald's office with questions or concerns, he would stop whatever he was doing and give me his full attention.

Admit Your Mistakes

If you are wrong, admit it. Managers are not perfect; they are human, and showing human qualities and frailties is a plus in any organization. Good leaders always take responsibility for their errors and never blame personal mistakes on others on their team. Demonstrating that errors are an expected part of the experience allows employees to feel less threatened, and they have more respect for you as a leader. Your team learns that honestly admitting mistakes is the best way to strategize and come up with effective remedies or solutions.

Listen in a Way That Encourages Employees to Talk to You

Management intimidates many employees, so good listening skills are extremely important to promote honesty and open communication among your team. Make sure you listen whenever an employee needs to share, and show them that you are willing to listen by stopping whatever you are doing and giving them your full attention. Don't prepare your answer while they are talking. Let them finish their dialog, then think about what they have said, and if you need time to provide feedback, ask them if you can get back to them. If you do have a ready answer, you can tell them after they finish talking. Let people know that they are important and worthy of your time, and don't be too busy to listen.

Be Clear in Your Requests

It is your responsibility to communicate effectively to your team, so that they will be certain what you need from them. Clear direction will enable the job to get done efficiently and with less interruptions or confusion. After delivering a message, always ask if everything was understood or if they need further explanation or clarification. No one likes to be given poor direction or misleading information because it makes the job infinitely more difficult and increases the chance for error. Remember, your job is facilitating work, so communicate it.

Stand behind Your People

Supporting your team is a critical piece of good leadership and can be very challenging, especially when an employee fails. Nevertheless, it is important that your team feels they have your support in any and all circumstances. If workers feel that you will not stand up for them, then you have failed as a leader. I knew a manager who was not seen as an effective leader because he would sacrifice his team whenever something went wrong, even if it meant he needed to lie. If one of the team members made a mistake, he left them standing alone in the cold, which is exactly the opposite of how a good leader should handle this type of situation.

Be a Good Communicator

Employees respect a leader who can articulate what they did wrong, without blaming them. Many times, mistakes can be made because of unclear direction, so look to yourself first for possible reasons for any mishap, and admit your mistake and responsibility first before explaining theirs.

Employees look for a leader who is not secretive and who will pass on important pieces of information about the company. Honest communication halts the rumor mill and false speculations that are typical in any organization, and promotes trust among your team.

LEADERS

I have been surrounded by all types of leaders my entire life. I have participated in team sports since childhood, which provided many opportunities to lead and be led, as did my many years in academic institutions. I have always had a healthy respect for effective leadership, and I know that people like to be led, to have structure and discipline, and to be asked to do things that have importance and relevance.

Poor Lean leadership definitely results in lack of motivation, poor performance, high absenteeism, and, ultimately, high employee turnover. They are easily recognizable because they have all or some of the following characteristics: they practice negative reinforcement, they are focused on their own personal needs rather than the professional needs of their team, they are pessimistic rather than positive, they are poor listeners, they are lazy or lack motivation, they are stubborn or closed to new ideas, they are slow to adapt to change, they are blamers rather than responsibility takers, they provide bad direction or unclear direction, they have no idea who their people are, they are secretive, they are never available, their door is always closed, they fear failure, they do not stand behind their people, they have difficulty developing their employees, they exercise leadership by control, manipulation, and coercion. Not helpful to successfully engage people in Lean.

Effective Lean leadership is not based on control, coercion, and manipulation. Lean leaders are focused on the future rather than the past. Lean leaders gain respect by their ability to inspire others to work toward specific goals. Effective leaders help others become better people and create workplaces that attract good individuals and keep them happy and motivated toward excellence.

The first step in being a successful manager is to admit that you don't have all the answers. Admitting that you are not all-knowing gains the respect of your employees, along with their trust. Being realistic is also a positive characteristic. Realize that it just isn't possible for you or your team to solve every problem that exists, and know how to establish boundaries for yourself and your organization. Be yourself, be authentic at all times, and remember that any failures will be forgiven if you are honest and always try to do the right thing.

Many of today's leaders lack fundamental leadership skills. I hold the personal belief that leaders are born and managers are trained. Businesses have made gross mistakes by trying to turn managers into leaders, which is not always possible because leadership characteristics cannot always be learned. The major distinction between leaders and managers is the fact that leaders recognize the value of people, while managers drive business. Managers are listened to, but leaders are followed.

Poor leaders create a variety of problems for a company. Often, their actions result in creating overtime in a work environment that is already very unpleasant because they simply don't understand the problem or how to effectively solve it. Many poor leaders are unable to identify skill sets in their organization; therefore, work is assigned to the wrong persons, which increases poor morale, poor production, and eventually lost revenues. Lost revenues lead to downsizing and layoffs, which then prompts poor leaders to burden the still employed with unreasonable workloads and more responsibilities that do not match their skills, education, or training.

Assigning work to employees who are unable to perform that work creates overtime as well as confusion and anxiety for those trying to accomplish the company goals in unrealistic situations.

Work should be distributed based on the skill level of employees, and leaders should develop plans to provide the appropriate tools and resources to those on the team who need more development. This is not an exact science and not an easy task, especially when companies are always focused on the bottom line. However, good leaders are attuned to their people and know how to keep everyone doing the right thing for them personally as well as for the company. Effective leaders know that every employee has something to offer and acknowledge their responsibility to utilize that talent with projects and assignments that correlate with their ability.

I encourage all of you to challenge your managers to become better at what they do. We as leaders need to realize that people are the number 1 asset in a company. Stockholders do not show up to the company to work, the employees do. Being a Lean leader requires very good balancing skills because we have a tremendous amount of responsibility, not only to improving the bottom line, but also in the manner that we improve it. However, through ethical and sane leadership, we can help a company grow.

How you approach your people and help them embrace Lean and change starts with identifying how you treat them. Lean leadership is nearly impossible to teach, but I hope I have shed some light on the obvious. I want your journeys to be successful. Lead as needed.

NEED AN ROI

Avoid thinking about Lean only as a return on investment. Lean is not always about cost and the constant analyzing of its ROI. Lean is a business model, which is about developing the right type of processes, changing culture, and improving organizational performance. In Chapter 1 I wrote about becoming a metric-driven leader. You should; you have a business to run. However, being successful at achieving a financial return from Lean is also about improving elements of the business that don't necessarily bring you a metric return, or at least in the short term. If you get caught up in cost, you will not get the total buy-in and following of your people. So, you need to balance both. The problem with getting caught up in the financial return from Lean, at least in the beginning, is that you won't see it immediately. To be honest, as a Lean leader, you need to change the way you think about the financial gains from Lean. Opportunity cost and better cash flow is generally what occurs *after* Lean improvements. For instance, reducing labor costs is often common in Lean; but unless the people are released from the company, the labor costs still exist. Redeployment of the people is often performed and this labor becomes opportunity costs. If a production line before improvements required 4 people working to make 100 products a day, and now the process needs 2 people making the same 100 products, you have made it more efficient simply because you are better utilizing your people's time. The labor inputs are reduced and you are achieving the same result. Unit cost of the products decreases as well. But still, if those two other operators are now working somewhere else, you have only diverted that labor cost.

You generally see real cost savings from inventory reduction, not only in the short term, but ongoing. Leading inventory reduction projects involve the establishment of new maximum and minimum levels for reordering. Often through this exercise companies see that they are spending way more than necessary on supplies and parts as compared to what they are actually using. As they challenge maximum levels, they shrink the gap from what is bought and used. Instantly, buying is turned off until the existing inventory is "bled" down to new levels. This represents a savings to the current inventory spending. As time goes on, you now have new buying habits that keep inventory low, which, in essence, is less spending than before.

Labor is much harder to see. If your organization is running at a high level of overtime to meet customers' orders and, through Lean efforts, you streamline the operation allowing the same people to produce more, then you simply reduced or eliminated the need for overtime. This allows you to operate within the correct labor standards.

However, there are situations where an implementation of Lean is needed to keep the company open. I have been part of Lean transformations that were necessary to avoid sending all the work overseas. After some aggressive implementations and major efficiency gains, the company did lay off all the temporary workers. These temporary workers were employed due to the high level of waste in the production processes, and the only way the organization could meet delivery dates was to add excessive resources. By no means am I an advocate of job loss, even in the temporary ranks, but their services were no longer needed. The business was able to now work within its labor standards and become profitable again. This reduction in workforce was truly a cost savings. Just remember that not all Lean journeys are for growth. Sometimes it is for survival.

Nevertheless, don't get caught up in the ROI too quickly because it takes time to achieve it. Lean provides a foundation for better cash flow, opportunity cost, and smarter buying habits that yield better margins at year end.

Lean is a business model for change and continuous improvement, and you as a Lean leader will invest time and money into your journey that won't always yield a metric improvement. Here are some examples:

- Training
- Team building
- People development
- Strategic planning

Training

It will be extremely difficult to implement a new business strategy and have successful implementations without an understanding of the tools within Lean. If you hire a trainer, take an online course, or develop your own curriculum through books, money and time will be spent during this phase. Involve as many people in the company as you can. Arming your people with the needed information will show support from your side, but prepare them for the actual changes. More importantly, you must move ahead with implementation so the employees don't think the training was a waste of time.

Team Building

You will find that, through your Lean journey, you will unite your employees and create a great team-building atmosphere. Lean projects involve cross-functional teams performing the implementation. Often, departments that never really worked together are now placed on teams to challenge current practices, remove waste, and improve the process. I have witnessed time and time again where teams involve CFOs, presidents, operators, maintenance, purchasing, plant managers, supervisors, and office staff. Everyone working side by side learning from each other and helping the company is a great thing. I cannot place a financial value on this, but it is extremely beneficial to your success with Lean.

People Development

We are always being told by financial advisors to make our money work for us and turn that investment into extremely valuable assets. Your employees are no different. As you progress with continuous improvement, you slowly develop your staff into valuable assets for the company. Many employees, including executive leadership, can get into a stagnant job where it is the same thing over and over. Developing your people is a vital component of Lean.

Strategic Planning

A fundamental element of Lean is the creation of a Lean strategy, as described in Chapter 1. To be honest, several large companies I have worked with struggled with this policy making. As a Lean leader, you will find yourself developing in a very progressive manner. As you develop a mind set for change and dealing with resistance, you also will find your policy deployment skills sharpen. Successful Lean implementations always come back to strategic planning, deployment of the policy, and project management.

WORLD-CLASS PRODUCTS NEED WORLD-CLASS PROCESSES

Most companies in industry feel that their product or service is best in class. Small business owners are at the top of this as they possess the entrepreneurial spirit. They have a close attachment to what they make or what they offer.

Take a moment and think about the processes involved in producing your product or service. Do you market your product as the best or one of the best? Great quality, seamless application, best customer service, longevity and reliability? Now, step back and visualize the environment in which it is made. World-class products or services need world-class processes—period! Would you give your customers a tour of your facility, or take them to a showroom or controlled R/D lab to show them the product. As you lead your organization down the trail of continuous improvement, you have to develop a process-oriented mind set. Every product, service, or administrative function has a process. Having a world-class process is necessary to back your claims of a world-class product or service.

YOU ARE THE CREATOR OF YOUR BUSINESS REALITY

I am a firm believer in the concept of "ask and it is given." The mind is a very powerful thing and you must go into your Lean endeavors with a positive attitude and mentality of success. If you create negativity around Lean, don't empower your people, don't train them properly, and even downplay the success of continuous improvement, you will fail. Call it new age, I don't care. You have the power to create the environment you want for success with Lean. Embarking on the journey in itself is bold, but forge ahead with positive energy.

I have seen it plenty of times when comparing successful Lean companies and those that are not as successful. It always falls back on leadership commitment and attitude. Grumpy business owners and executive leaders have very little or no success with Lean. They are constantly questioning the philosophy and find ways to literally sabotage the journey. Strong and result-driven Lean journeys are seen with company leaders who have the complete opposite approach. Positive and forward-thinking leaders will make it happen. Yes, they come upon obstacles and resistance, but they break through them and help lead their companies down a path of success.

You must be ready to lead the Lean journey. It will change you and your employees in a way that most business philosophies cannot. You will learn a lot about yourself and your employees as it progresses, and I also believe that once you are infected with the Lean bug, you will never get rid of it. In a good way, of course.

3

The Psychology of Waste

As mentioned in the Introduction, waste exists at every level of the company and it affects people in a variety of ways. It often can create a feeling of comfort at the moment of its occurrence because it can act as a nice buffer to problems. Unfortunately, this feel-good state from waste will have determinate effects on the long-term health and growth of your company.

As you know now, many organizational leaders struggle to admit the need to improve their business, find it difficult to approve the financial investment in Lean, and sometimes never realize that it is a business model rather than a program. Thus, waste stays in the operational processes and eats away at cost, quality, and delivery. As you read this chapter, I want you to think about your organization and find where the psychology of each waste is affecting your people. More importantly, I hope it encourages you to take action and reduce the waste. Keep in mind, though, that companies come in various shapes and forms, and the products and services they offer encompass a wide spectrum. Always remember that waste is everywhere in a company. Therefore, if you work, for example, in an office, manufacturing, warehouse, food processing, or auto repair and auto body repair, your organization is still producing something. It is either a product or a service product. Below are examples for all types.

PSYCHOLOGY OF OVERPRODUCTION

Overproduction is the mother of all wastes. It is by far the most common in most industries, and it has the ability to create all the seven other wastes. It has a unique feel and impact on your culture. Overproduction is at the top of the "feel-good" list. In manufacturing, the transformation of material and parts into a finished good is the whole purpose of its being. The more finished goods made, the more revenue that is generated. The problem that exists is that the finished good is valuable to the company if it actually has a buyer. Not a buyer in theory, but an actual order that is from an external customer. When trying to reduce the gap between your inputs and outputs, it is imperative that once the product is complete, it leaves the facility at a pace that brings in revenue. The longer it sits around, the longer it will take to be paid.

The first feel-good comes from those companies that really enjoy piling up finished goods. Oh, yes. Fill those racks… fill to the sky! Those finished products look very nice and provide a false interpretation of productivity and the fruits of your labor.

If you are a warehouse or distribution center, however, then your level of finished goods should be at a level that can meet the larger number of customer orders from various places. That is different. I have been on many plant and company tours in my career. Often the conversation that comes up when walking with plant managers or vice presidents of operations is the need to cut cost or improve cash flow. Of course, from a Lean perspective, we want to control costs, but many companies that embark on Lean are looking only at the financial return.

As I walk the aisles of a plant and I see the rows and rows of finished product ready for sale, I asked the tough question: Are you a manufacturer or a warehouse? This by no means implies that the goals should be of no finished goods whatsoever; rather, it means a reduced amount that better represents the buying patterns of your customers.

I had a tour from a plant manager once, discussing this exact subject and he understood my question and agreed that a lot of the company's money was tied up in finished goods. However, before the Lean bug had hit him, those finished products felt good. It creates a feeling of completion and that the company's inputs have returned a sellable output. The irony here is that no money has come in or they are not truly getting anything *out* of it.

The more interesting part of overproducing finished goods is that it provides a poor measurement of productivity; thus this measurement is not accurate. For instance, in companies that have a lot of overproduction where they are simply filling the racks and shelves, they have no sense of targets. "Oh, boy, we are really behind," or "We are so ahead of schedule." Based on what? It is based on feel. Not every process can be making products only to firm orders. Some can, but often there is a mix of "to order" and "to stock." Stay with me on this example.

Thus, in essence, the organization is measuring its ability to simply over-produce. The inputs are there and the output (the finished product) is there; however, there is no achievement of what you really want—revenue. It may come at some point, but this indication of productivity is false. The best measurement is the inputs and the outputs of something sold; labor dollars per units sold. This is the best measurement of your inputs to your outputs.

Moving away from the finished goods example, we now venture into the actual production areas, such as assembly, fabrication, welding, brazing, painting, etc. This is where overproduction has the most profound impact on your culture. Overproduction in the production process is an indication of a variety of problems. Without clear goals of output, set production levels, or pace, machine and manual working operators will overproduce. There may be variation in the process where people are not following the same steps and have consistent cycle times. Workloads between assemblers are not balanced. If operator A has 15 minutes worth of work and the next person, operator B, has 25 minutes worth of work, then operator A will more likely overproduce and accumulate partially completed product (work-in-process (WIP)) between the two. First of all, operator B will probably begin to feel overwhelmed and possibly make a mistake trying to hurry. A non-Lean-thinking supervisor will have the impression that operator B is slow and that operator A is quite the ace worker. It does not matter if operator A builds product out the door and down the street; nothing is being accomplished at the end of the process for the customer. All that inputted labor and material is for naught.

The operator now begins to feel very good about those accomplishments as there is a nice pile of product sitting there that represents the effort. And, finally, what might happen is that operator A may now begin to slow down or even leave the work area because there is a pile of work sitting. Hence, the feeling from operator A is, "I am done; here you go." As a leader, this would be doing your production workers a major disservice with this false sense of completion.

The first non-Lean thought is that something is wrong with the pace of operator B, when, it fact, it's the process. People are only as successful as the process in which they are placed to work. The process in this example is conducive to overproduction and waiting, and people who work in it will do precisely that.

Overproduction also can have another psychological effect on people. As WIP begins to fill up everywhere, the work area becomes more cluttered and disruptive. Piles become obstacles, physical visibility is reduced, and the sense of "on time" is lost. As product begins to fill the empty spaces, you lose your sense of urgency in the process. You don't lose speed, because speed does nothing. You lose the customer's needed sense of urgency to produce a product on its scheduled delivery. Production workers tend to slow down as they become congested in their environment. As you begin to challenge this WIP and begin to reduce it significantly, not only will you see output increase, but the thinking of the people changes to that of productive soldiers. Less stuff in their way of the finish line really helps create that needed sense of urgency—not only to the external customer, but the immediate internal customer as well.

We continue on yet again into the welding, painting, and fabrication departments. Any environment where manual or machine setups are necessary, the potential for overproduction begins to grow. In these environments, there must be a balance between reducing overproduction of the fabrication parts and utilizing the machines properly to achieve a return on investment (ROI).

The first psychological battle that begins is that of traditional manufacturing. Purchasing of equipment and machines usually is a long process. The machines' capabilities and cost are highly evaluated to ensure it can produce the product needed to generate sufficient revenue to pay off the machine' and create sustainable profitability. The question of ROI comes up when evaluating which machine to buy. How many parts are produced and sold (*sold* being the key word here) before the machine is paid in full? Companies often forget the *sold* part of this question.

Setup times should be an integral part in this equation because the longer the setup, the longer the time until value-added parts are made. Often with long setup times, manufacturers tend to produce large lots of products because the thought of having to change over again is too painful. In a schedule-driven environment where companies are producing far ahead of schedule, the operator may opt to set up the machine, run part A, and then run all the part As in, let's say, a two-week schedule. Changing over is not an option. The operator will simply make all the needed parts of that setup before changing to another lot. The psychological effect here is that this operator or supervisor feels that the investment in the machine is being maximized: "Keep those parts coming." With downtime being the moment of nonproduction, keeping the machine running creates another false impression of success or productivity.

Unfortunately, the next lot of parts in line for a different customer simply waits and creates delays all the way back down the line. Reducing setup times is the key, not overproducing. You have to deal with all the inventory that accumulated, and you better hope the parts are correct, handled safely, the customer does not change its order, and no engineering design change is coming.

We can construct a rather silly analogy with this. In an office environment, the machines that help the office workers perform their work are items such as copiers, printers, computers, servers, and scanners, to name a few. Just because the copy machine can copy 10,000 pages does mean you need to copy that many. Yes, you may need to copy that particular work order or office memo next week or the week after, but there may be better use of your time, and that of the copier.

It reminds me of the *Toast Kaizen* video by GBMP (Greater Boston Manufacturing Partnership) and narrated by Bruce Hamilton. In the video, Hamilton is explaining the overproduction of a toaster. He is making toast for his wife to illustrate how waste extended the lead time to the successful completion of a snack for his wife. In the beginning of the video, he proceeds to make four pieces of toast. When he is done, he prepares to hand it to his wife and she explains she wanted raisin toast instead of regular toast. He throws away the toast and improves the communication and process. In the new and improved manner, he asks her how many and what kind of bread. She answers that she wants two pieces of raisin toast. His later explanation of reducing overproduction on a machine was that just because the toaster holds four pieces of bread does not mean he needed to make four pieces. Yes, he did not utilize the whole toaster and its intended quantity usage, but he did not create scrap, inventory, and waste his time on something not needed.

In the auto collision repair industry, we use a similar model of overproduction and false sense of flow. The auto collision repair industry is one of the most diverse environments I have ever seen. They have little or no visibility of the work coming in from one day to the next. Often, shop owners come to work and there are two or three cars in the parking lot. There could be a two-door 2006 Honda Civic, a 2001 Chrysler Town and Country van with a sliding door, and there could be a 1999 Ford Ranger 4 × 4 pickup. Oh, and all three are of a different color with major differences in the level of damage. My point is that this type of dilemma lends itself to different needs through the repair process.

In a typical auto collision repair shop, there are about five definable stages. First, there is a dismantling process that is performed to establish the level of repairs and the type and quantity of replacement parts. Next, there is a body or preparation area that begins the repair of the panels, bumpers, doors, etc. Next is paint, then reassemble where new and repaired parts are installed, and often the shop will clean the car for the owner in a detail area.

Depending on the level of damage, model, car, and year, the repair process will be different in each process it enters. Thus, the cycle time in each process is going to be different. Like an assembly line, this opens up the high chance of overproduction. It has the same effect on technicians as it does on assembly line workers: too fast, too slow, slow down, walk away, "I'm done, here you go."

Once again, the same feeling can happen in the body shop's office, or in a bank, an insurance company, or any administrative environment. Hence, estimates, work orders, loan applications, insurance quotes pile up and stay in piles until someone is ready for them. In our body shop example, the paperwork flow is mostly in estimates. Estimates are frantically being completed for the shop, but the shop is unable to use them. Pressure begins to build, creating unneeded stress on the technicians to hurry up. Therefore, if the repair process they are given to work on is unbalanced, contains variation between workers, and capacity is maxed out, a sense of failure begins to sink in.

One administrative overproduction example I recently witnessed was during a visit with a customer in the South. The project was to analyze the work order process in production control as orders were sent from customer service. Production control was responsible for putting together work order information in regard to machine setup, parts needed, color, labor hours, etc. The paperwork process was long and tedious, and there was a lot of extra checking, verifying, and overprocessing. Most of the team performing the evaluation definitely saw the redundancy in the work order process and began to come up with ways to reduce it.

As we were going along, I witnessed the start of a work order as an employee began to input machine requirements. Then she entered the delivery date: December 18, 2010. Keep in mind, this analysis was being conducted in July 2010. This was work order overproduction. All work orders were completed in the office, regardless of the completion or delivery date. In fact, the department spent nearly 40% of the time processing work orders not due for delivery for more than six months. Production supervisors were always complaining about down machines due to no work order. Often, the work orders would come to the factory floor in the wrong order.

The perception was that production control was slow and it created animosity between the office and production. I spoke with the people in production control and they felt overworked, behind schedule, and stressed. So, overproduction is not always the quantity of work, but the order in which things are done. Once we set a guideline on when to start orders based on ship dates, everything changed from a psychological perspective. That was all we did in the first pass. The second pass was to eliminate the redundancy.

As you can see, overproduction is brutal on the mind and perception of work. Even though you will never completely eliminate overproduction, identifying overproduction and reducing it will have a positive impact on your culture and provide a real sense of completion of your staff's hard work.

PSYCHOLOGY OF OVERPROCESSING

As you may be starting to realize, waste is a feel-good in business. It is in every process and it simply is a buffer to real problems. The waste of overprocessing is the act of redundant effort or extra steps. As mentioned in the Introduction, it also surfaces in processes where defining the completion or endpoint is difficult. In business, we as leaders and owners come across situations where we as a company made a mistake on either a product or in the delivery of a service that created a problem internally or for the customer. It may not have necessarily caused physical damage, but a situation where we screwed up, pure and simple.

For instance, in a warehouse where there are a large number of products and orders processed every day, the environment is very order driven. By nature, there is not a lot of overproduction from a perspective of quantity. You order one type of blender; they won't send you two just because they could pull two off the shelf at the same time. Pulling orders in the wrong sequence is overproduction in that environment.

In a non-Lean-thinking warehouse, rather than reduce the waste in the process, triple checks, added steps, and reverifications are put in place to ensure that this problem never happens. This process takes more time or more people, and the problem may still exist.

This is an example of how overprocessing can create a feeling of false control of the work order picking process: "We double-check everything." Stress is created within the culture because the warehouse staff never want that particular mistake to ever happen again; and, if the extra steps are not taken on a given order, then the employees may become afraid that a potential defect will be sent out.

Overprocessing does not resolve anything. It provides a band-aid for mistakes and merely hides the inefficiency. Administrative functions contain all types of overprocessing. Do you copy people on e-mails who really do not need to be copied? Then, do these e-mails go into special inboxes to be saved for no future purpose? Offices often hold on to e-mails like a lifeline as way to shield them from getting into trouble. Is information saved in e-mails, hard drives, external drives, discs, flash drives, and in servers when really one or two is sufficient for security? Overprocessing simply is another feel-good waste. Some level may be "necessary waste," but you probably see my point.

Fabrication departments, food processing, painting processes, and auto collision repair and machine shops have an opportunity to reduce overprocessing like anyone else. Redundancy does exist, but the type of overprocessing I am referring to is when a process has an unclear completion point.

Ask yourself this question: How do you define clean, sanded, buffed, deburred, painted? It's tougher than you think. You will define completion for some examples differently than the next. Here is a fun game to play with your employees. Place four people at a table. Hand each one a 6-inch piece of 2×4 of the same species of wood. Then, place on the table sandpaper, an electric sander, and another piece of sandpaper still in its package. Tell them that their job is to sand the wood until it is smooth and when they are done, have them raise their hand for you to perform a quality check. Make sure you have a clock or stopwatch to time the exercise. And then tell them to start. Smooth is hard to define and it will become very apparent. First of all, each person will probably start with the sandpaper given to them and they will all stop at different points, checking and rechecking their work. Maybe, one person will sand for one minute and stop, raise her hand, and ask for a quality check. You may be satisfied with her piece of wood and tell her she is done. She will probably be very excited about her accomplishment and goof around with her co-workers in the exercise. One person may stop sanding and decide to open the package of sandpaper of a different size and continue to sand even more. Another person may waste no time at all and grab the electric sander and proceed to use it to complete the job. The fourth person may just sit there with the original sandpaper and be perfectly content that he will have his block of wood sanded effectively. All four participants will finish at a different time. All four pieces of wood will appear smooth, but will probably have different levels of abrasiveness.

What also might happen is that the person who used the electric sander may run his finger over the piece of wood that was completed first and proceed to sand it more. The woman who stopped first may be a little disgruntled about this because she was happy with her work, but is offended that this guy was not. There may be some arguments over whose wood is the smoothest and, really, what you, as a quality person, define as smooth. Maybe some level of resanding is needed by two of the individuals to equal that of the others.

I have tried this simulation in my training and it is quite interesting to observe the psychology that goes on. The amount of resanding was different from one group to the next based on their interpretation of smooth. Each person performed the work with different tools and in very different cycle times. There were minor debates and arguments between the workers on the definition of "sanded." This type of overprocessing in real work situations creates confusion for your people and also can prolong the training time needed for new employees.

PSYCHOLOGY OF MOTION AND TRANSPORTATION

I am sure you or someone else you know has used the term: "Don't work fast, work smart." As old as the saying may be, I use it all the time when speaking about Lean or productivity. Speed gets you nowhere and can actually slow you down in the long term. The waste of motion and transportation can equally hide the concept of completion or success and make people work faster.

Motion is the wasted movement of people walking around the office, the production floor, warehouse, or a repair shop, for example. As people walk away from their work, the time it takes to complete the work is lengthened. Now, keep in mind, I am aware of the importance of breaking the monotony of work, circulating production workers, and the overall wellness of people. However, in the eyes of the customer, motion is 100% nonvalue added.

On a production line, operators are either performing value-added or nonvalue-added work. Value-added work is the steps that change the fit, form, or function of the product. Simply put, installing parts into a partially completed product. In an office, it could be filling out the work order on a computer with vital information about the part to be made or picked from the warehouse. In a bank, it is the work spent filling in the steps required to verify credit, and in a body shop, the painting of a bumper. All of these examples are changing the fit, form, or function of product or service product on which a person is working.

As the work is closer to completion, the workers in the examples above begins to feel a sense of accomplishment or feels they are working hard. Motion has a similar effect of working hard, but the sense of accomplishment is once again false.

In a production line, as operators are performing value-added work, at certain points they may have to leave their work area to find tools, supplies, people, information, and material. These actions take them away from their work and then create a negative impact on delivery and productivity. As each day, week, and month goes by, more and more time is consumed in motion. Concurrently, as each day, week, and month goes by, production numbers begin to slip, deliveries are late, and productivity decreases.

As an outside observer watching this person walk here and there, then build, then walk, then build, the perception may be that the workers is, well, working hard. You are darn right they are, but less and less is getting completed. Motion can make your workers feel overworked, especially in companies that manufacture product or repair cars.

Many times in my travels I come across production workers who work in a process where they are constantly walking around looking for the things they need to work. When asked about how they feel about their work environment, I hear things such as, "We do what we have to do to get the jobs done." "We make it happen." "We work hard." Well, all three responses are correct, but really, these comments tell me there is dysfunction in the process. Plus, it takes a lot of effort and can be tiring. As a leader of a manufacturing or repair business where tangible products are completed in some form or fashion, wasted motion can have a long-term psychological effect on the workers. Motion simply takes the finish line for each process and extends it farther out, giving your people false impressions of success and increasing the time they can continue conducting value-added work.

It's almost like comparing the 100-meter dash and the 100-meter hurdles. Which one takes the longest time? Which one has the most obstacles in front of the runner? Wasted motion is like dealing with hurdles. As these hurdles are reduced, your runners (workers) are able to the see the end sooner than before; and as each hurdle comes down, they will begin to feel that true sense of completion that each production and repair person would like to have.

Motion is often unseen in traditional accounting practices. It is here where Lean accounting practices take a different approach. Most production-related operations or even auto body repair view workers as either direct or indirect labor. Direct labor is defined as workers who add value, build products, repair products, or change fit, form, or function in some way. Direct labor is the key ingredient in unit costing methods when establishing standards. So, the better the use of direct labor, the more profitable the product offering is. Indirect labor is work that does not add value, is considered overhead, and must be absorbed somehow into the company as an expense. As an example, a technician repairing a dented door is direct and the person delivering material is indirect labor.

If technicians have to leave the work area to find a tool or parts needed to perform the repair, that actual action has turned them into indirect labor. The accounting books won't show it, but that is precisely what it is. Thus, motion is giving you a false sense of job-costing as well.

Transportation is defined differently than motion, but the effects on the mind are basically the same. Transportation is the movement of products or information within an organization and it often requires a person or a machine to perform the movement. I would like to take an interesting approach to transportation in a way that most of you can relate to.

Think about your office, warehouse, or production facility. Is the office laid out based on departments? Is accounting, marketing, customer service, graphics, accounts payable and receivable, or production control, as examples, spread out? In traditional manufacturing facilities, we call this process-based production. Departments, such as welding, fabrication, painting, assembly, packaging, or grinding area, are all in separate locations. Regardless of the reasons why, this type of environment creates a reduced sense of teamwork, lost visibility of the "internal customer," and plenty of transportation. Once work is completed in an area or department, it is simply pushed into the next area for processing. It may be from sales to marketing, graphics to production, welding to assembly, or customer service to accounting. It goes back to what I wrote about overproduction: "I am done; here you go." But, of course, with regard to transportation, movement is now needed and it is this space between processes that reduces your need for overall teamwork. In the eyes of your customers, they don't care; they want their product or service product in a timely manner, of good quality, and at a price that is acceptable.

In manufacturing, process-based production is at the core of the creation of other waste. Piles of product are made (overproduction), stored (inventory), moved (transportation), stored (inventory), and then finally worked on. Thus, the individual psychological effects of the wastes begin to multiply.

PSYCHOLOGY OF INVENTORY

I expect some level of resistance from my point of view here, but I encourage you to be open minded about the psychology of inventory. Before delving into the Economic Order Quality (EOQ) model and the waste of inventory, I want to discuss the concept of hoarding and the need to store everything. There is a very unique mentality behind people's attachment to stuff they do not own. I have witnessed this in offices, manufacturing, warehouses, and many other processes. Workers get personally connected to things in their work environment. Supplies of various kinds are needed and the list is quite long, depending on what the company does. In an office, there are desks, cubicles, chairs, filing cabinets, folders, paper, pens, staples, staplers, highlighters, etc. In manufacturing, you find masking tape, rags, cleaners, gloves, masks, eye protection, ear plugs, sandpaper, oil, etc. Company supplies can be one of the highest costs in an organization, and this cost is not necessarily accounted for in in-product cost. Usually, the tangible items that actually ship with the product or service product are part of unit cost. In a non-Lean environment where there is little control over company supplies, employees begin to fill their space with escalated quantities of these consumable items.

One improvement tool within Lean is called 5S (sort, set in order, shine, standardize, sustain), which I describe in later chapters. Part of 5S implementations is sorting through the work area and removing things that are not needed. When conducting this sorting phase, the 5S team reveals that the workers have been hoarding supplies. For a production worker, it is the fear of running out of that favorite eye protection or gloves. In some cases, if there is a problem with workers taking each other's supplies, the perceived solution is to take as much as possible from the stockroom and hide it in lockers, toolboxes, and lunch boxes. Office workers fill their desks and cabinets with large amounts of pens, notepads, ink, toner, highlighters, and other items.

When organizations have no process in place to control the use of supplies, employees will begin to fill their work areas, and this costs a lot of money. It is about bridging the gap between what is used and what is bought. These excessive supplies are an example of inventory and provides a level of security for those hoarding.

My examples may appear small in nature, but think of those companies that have 250,000 square feet of manufacturing space, multiple floors of offices, or even multiple buildings. A significant amount of company money is tied up in this poor use of supplies, and I recommend a serious look at the problem. There are bigger problems at hand creating this excessive use of supplies that must be fixed. You will find resistance in your first pass at reducing this inventory, but it will raise the issue to the surface.

One problem that I see with hoarding is that the company allows it to dictate its buying habits. It is here we begin to see this large gap between usage and what is purchased. Part of the EOQ model is to manage material costs in the product or service product to achieve a greater margin on unit cost. Regardless, if it is actual parts or material or supplies needed to run the day-to-day operation, many traditional purchasing philosophies that follow EOQ models try to buy in larger quantities to reduce the unit cost from the supplier and, hence, affect the finished good's profit margin. A long-term and ongoing goal of Lean is to constantly challenge how inventory is perceived and the manner in which it is purchased. Reducing inventory quantities and costs will improve your business in so many ways. But first, let's look at how it is hurting you and how it, yet again, affects the psyche of your people.

The lure of bulk discounts has been a mainstay in the United States and many other countries. With the emergence of places such as Sam's Club, Costco, and other bulk retailers, we as consumers are attracted to the concept of "buy more, save more." In many cases, it is a smart decision to buy things in large quantities to save in the long run. The model would not necessarily work on perishables such as food due to expiration dates. In manufacturing, the same buying model of bulk discounts is very common with the intension of reducing unit cost. The concept of "buy more, save more" is quite often an illusion of savings and can have long-term negative impacts on your company's bottom line. What we are trying to accomplish through the reduction of wasteful inventory is to shrink the gap between what is used and what is purchased. Often the gap is very wide.

There is a wonderful feeling for the buyer and warehouse person when looking at the large quantities of inventory and supplies. It is quite a win because, if they were to get a deal on particular parts or supplies, they would feel that they have saved the company a significant amount of money. Keep in mind that the ultimate buying decision coming from the top is to find the best deal per unit price, so you can't blame the buyer.

So, let's say a buyer receives a phone call from a vendor that there will be a unit price reduction of 15% if the buyer purchases a six-month supply. A quick look at the standard material cost in manufacturing resource planning (MRP) suggests a great saving per unit that will result in an increase in unit profitability. The decision is made and the six-month supply is on the way. Now, keep in mind that within 30 days or so, a lot of parts need to be paid for with company funds.

A few weeks go by and the delivery company shows up with a trailer full of discounted parts. Now the fun begins. First, a forklift is needed to unload the multitude of pallets. Remember forklifts require maintenance, driver certifications, and insurance. Not implying a forklift was not already present, but a transportation device is needed to handle inventory. Someone needs to drive this forklift (labor), and have certification renewed as needed. The forklift driver begins to move the pallets out of the trailer and into storage. So, make sure to have enough racking and shelving on hand to accommodate those pallets. Racking and shelving need floor space, one of our key shop floor metrics, and that floor space needs a facility. The inventory on the shelf must be controlled by some type of material management system and probably a handful of employees.

As inventory sits around, it becomes a target for damage by forklifts, lift trucks, pallet jacks, people, etc. Damage to valuable inventory quickly makes it invaluable. Also, there is a chance that there are some number of defective parts in the six-month supply and it becomes visible two months later. Finally, let's hope there is not an engineering change request looming that will make the discounted parts useless. And, when we compound this purchasing model over multiple parts and supplies, things add up quickly. All of sudden that 15% discount has just become a bill.

The psychological impact on the mind of the buyer or purchasing department is that they are saving the company money. It's really not one person's fault; it's just how it is in the company.

We now think of our plant manager walking the facility complaining about cash flow and floor space issues. This is how it happens and a reverse in your thinking about inventory is necessary because Lean accounting practices are concerned about many things, but, more importantly, the final profitability of the company. Yes, traditional accounting will show a unit cost reduction, but the overall plant/company cost reduction is not achieved.

A tremendous amount of effort is performed to control things we don't need or items of excess. Large amounts of time and money are spent in the storage of excessive inventory, supplies, and stuff. We as humans love our stuff. Look in your garage or attic; and even more frightening, hop in your car and head to the self-storage container you rent every month. This same model of buying in excess is just as prominent in the business world. Once that space runs out in the existing facility, companies then will spend money and rent outside storage. Thus, a vicious cycle begins.

From a Lean perspective, all this money tied up in excessive inventory and handling inventory may be better used somewhere in the company, and that is the change in thinking I want from you.

PSYCHOLOGY OF DEFECTS

Mistakes are going to happen in any business when people are part of the process. Most elements of a process are controllable, but some variation does exist. There are many components to a process to make it functional and produce acceptable quality. Most processes usually contain these key elements in some shape or form:

- Machine or equipment
- Power or other facility needs
- Tools
- Parts and materials
- Supplies
- Workstations
- Software and systems
- People

In an office environment, an accounting department has computers, printers, fax machines, calculators, spreadsheets, office supplies, paper, desks, chairs, accounting software, and bookkeepers and accountants. In the production environment, there are workbenches, drill presses, hand tools, power tools, parts, raw material, gloves, glue, fixtures, and production workers. Regardless of the process, the one component that has the most variation is people. Most of the other items in a process are mostly controllable. Machines and equipment can be calibrated and set for consistent output and quality. Parts can be made to the correct tolerances and quantities. Tools generally work with consistent performance, and software is pretty reliable. This is not to say that these items cannot break down after extended use or do not need general maintenance to keep them working properly; my point is that every day because humans are in the process there is variability.

It is important to "dial in" the controllable elements of your process so you are not relying on people to create the quality. Yes, it is their responsibility to be part of the process, but you must design the process to have the least amount of quality variation. As production or office processes begin to create errors and defects, that sense of accomplishment that you need begins to diminish. Negative managers are generally focusing on the people when quality problems are occurring and place blame on humans. People are only part of the process, and it has been my experience over the years that most quality problems occur from the controllable components of the process. Remember, people are only as successful as the process in which they are given to work. As people are placed into positions to fail and the leadership does not address the real problems, the workers begin to accept the mentality that they are mistake prone.

It is these same workers who probably know the problems with the machines, the server, the computer, the material, the tools, and no one is listening. Nothing feels worse than not completing the required work for the day due to defects and knowing you have no control over it. The workers keep working in waste, keep creating mistakes, and just work harder.

My point is that you must invest in the process to reduce variation. I don't mean variation from a Six Sigma* perspective, but provide a process that promotes success. Machines don't care if they make defects, and neither does an air tool. However, a person does and, if the process they work on encourages defects, then you are once again creating an environment where waste will occur.

---

* Six Sigma simply means a measure of quality that strives for near-perfection. It is a disciplined, date-driven approach and methodology for eliminating defects (driving toward six standard deviations between the mean and the nearest specification limit) in any process—from manufacturing to transactional and from product to service. (From www.isixsigma.com/index)

PSYCHOLOGY OF WAITING

People want to be successful and do a good job. I think we can all agree to that. There are a few employees in any company who are there for only a paycheck, but most of the time the employees want to feel they contribute and do the work they are intended to do.

The waste of waiting can slowly create a process of complacent workers. It also reduces the sense of urgency like overproduction. As you are probably beginning to see, waste creates waste. As a process is allowed to overproduce product, inventory is accumulated, and now transportation is needed to move the unneeded product. Delays in the process, or waiting, can come in many forms. Workers can wait on anything—parts, material, information, people, tools—and direction can be held up somewhere in the process causing waiting. It also is created when any process has imbalances in the work passed from one person to the next.

Waiting reduces the sense of urgency in the person sitting there. It also reduces your company's sense of teamwork. Inefficient processes with lots of waste can make your people not "trust" the system. For example, employees complain that

"Production control can never get us work orders on time."
"The estimator is always late on the estimate to the shop floor."
"Finance can never get us the approval when we need it."
"Welding is always behind schedule."

There is no sense of teamwork or coming together, and this is due to waiting. More importantly, it is a sign of a dysfunctional process or multiple dysfunctional processes all creating waste. Waiting also can decrease the desire to simply do work. Hurry up and wait! And, this happens again and again. If I was an employee in a process consumed by waiting, I can see how I would start thinking this as well. "Why should I do this since I know I am going to wait." It's almost like watching a football or basketball game on television when one team simply is not in sync with the other. Wrong plays are called. Players are not in positions where they are supposed to be. Communication between the players is not smooth, and everything is just out of synchronization. I can remember watching the Final Four of college basketball a few years ago when my favorite team, the University of North Carolina Tarheels, was playing in the national semifinals. From the very beginning of the game, they just were not working as a team. They were not executing the plays properly, communication was virtually absent, and players were just doing whatever they could to react to the opponents superb execution. There was no teamwork. The result in the first half was a huge double-digit score deficit that they were unable to break. My point is that you could just see the sense of failure and compliancy. The waste for sure was affecting their thinking. They lost the game, but all the players returned the next year and won the national title.

Waiting in a process will reduce teamwork and synergy in your company. As your processes become more in line, balanced, and communication flows more smoothly, you will instill a better sense of teamwork in the environment and waiting will decrease.

PSYCHOLOGY OF HUMAN POTENTIAL

As discussed in Chapter 2, Leading the Lean Journey, your mind set as a leader needs to change and adapt to this new business philosophy. How you treat people in a Lean journey can have a major impact on its success. There are a few items I would like to discuss here that all go back to making a process successful for people to work in.

"I'm Paid by the Hour"

I hate this quote, but I hear it a lot where waste of human potential is rampant. It usually is an indication that people are not being placed into positions that truly fit their skill sets. When looking at all the required jobs in a company, are people utilizing their expertise? Are welders being made to handle material, are shipping clerks on the assembly line, is the painter in the auto body shop writing estimates? Not implying that smaller companies need employees to wear multiple hats, but are they doing work they are not qualified to do? I have seen them slowly take on the mentality of "I don't care, I'm paid by the hour."

This environment now begins to reduce the sense of self-worth in the process because the worker is not doing work that fits their abilities. If it requires less skill or more skill, the person in this situation will lose faith in the placement process. You will eventually develop a workforce of "I don't care" thinkers.

No Investment in the People

Have you ever been in a situation where you're placed into a position to work and given no training or real orientation. This is extremely common in manufacturing (not so much in auto body), and we do see it in administrative functions. When manufacturers don't recognize the importance of preparing entry-level workers to make product, they are basically telling them they are not that important. It amazes me the lack of investment in production workers' success. When purchasing equipment, there generally is a capital appropriations process and engineering evaluations to ensure it is the right investment. Trial runs are performed, test pieces are made, and debugging software is conducted. Rightfully so, because it is this machine or equipment that will be used to produce value-added products to be sold that will generate revenue and profit for the company. So, why not the same approach to line workers?

Regardless of the level of work to be performed, the proper training and orientation can go a long way in helping the people contribute to a good quality product. This shows them they are important to the process and success of the company.

People Are Expendable

A few years ago, we worked with a contract welding company that produced a variety of customized parts and assemblies for a variety of industries. In the beginning, as with any new client, we evaluated the company's ability to drive change and make improvements as leaders. Work at the contract welder was manually intensive. Setups were extremely labor intensive due to a lot of motion and transportation of changing out fixtures. Material handling was a nightmare and backbreaking work. There were ergonomic issues everywhere, and it quickly became clear to me and one of my engineers that to get these production workers to buy into Lean, improvements to the safety and ergonomics would be needed.

As my team began to conduct our analysis of flow, cycle time, inventory, etc., my manufacturing engineer was taking notes on how to modify workstations, carts, fixtures, and many other items that would not only reduce setup and cycle times, but also make the work much easier. The changes also would make the area safer and the work less physical—a win for both sides from my perspective.

After concluding our analysis for the process, we presented to the leadership team our findings and recommendations for the first few Lean implementations. When my engineer presented his workstation design changes, the plant manager said, "Why would we need to do these improvements? I don't see this saving me money, but only costing me money." I sat back and held my breath and allowed my engineer to take this one himself. He replied, "These improvements will reduce cost in the long term and will ensure **buy-in** from the workers."

He also explained to the plant manager that sometimes it is these subtle changes to safety and ergonomics that makes the worker love Lean.

The plant managers response, "I don't need buy-in; there are 100 people out there who will gladly take this job if they (current workers) don't like it." Hence, people are expendable.

If this is the mentality and thinking of the leaders, then how do the workers feel about their self-worth? This Lean journey is now destined to fail because the people will never get onboard. If you are practicing wasted human potential at this level, then forget about Lean working for you.

PSYCHOLOGY OF WASTE: CONCLUSION

I am sure this chapter has you thinking now about your own processes and how waste is really affecting each employee. To be honest, if you are reading this book because you know it's time to change, then you are off to a great start. Before you can make improvements in your business, you must first admit that maybe the way you have operated in the past needs to change. I am not implying that through Lean you will eliminate all forms of waste and the brainwashing it creates. However, as you systematically reduce waste, you will create a much more positive environment to work in where people want to succeed, be recognized, and make your company even more competitive. If they stay stuck in a process that is not successful, then the product or service product you provide will show it ... and so will profits.

4

The Psychology of Dysfunction

World-class products need world-class processes. It confuses me a lot when I hear company leaders talk about the products they sell. Most companies feel that their product is one of the best in its field. Finding your product niche is critical to staying competitive, especially with the large variety of products available. I see world-class products everywhere, and we as consumers use plenty of them. From residential to commercial usage, the products we buy are getting better and better with ongoing changes to options and variety.

The road of product development can be very expensive. Thousands of man-hours are placed into new product development. Companies can spend millions of dollars in developing products, and the length of the projects can extend into years. Advertising campaigns are generated, product labeling, and other attributes of marketing take time, money, and people.

All of this is done to ensure that a world-class product is developed and will sell well for the company to generate solid revenue and healthy margins. Product is launched and thrown over the fence to production. Oops.

Your world-class product needs a world-class process in which to make it. Companies often forget this vital and necessary step during product development. Investing in the manufacturing process is critical to ensuring that production can effectively produce the product with good quality, reduced waste, and reduced cost. This chapter attempts to illustrate what happens when there is no investment in the manufacturing process. The chapter is divided into two sections:

- Misconception of Working Hard
- Living with Waste

MISCONCEPTION OF WORKING HARD

I briefly discussed this concept when explaining how the waste of overproduction creates a false sense of completion. Processes with a high level of waste, as a whole, force people to "work hard" to get their jobs done. A side effect of working hard is working fast, and the two together can negatively impact the way your employees gauge their performance.

Waste takes people away from the work they were hired to do. Production workers, office employees, auto body technicians, bank tellers, and others, all are hired to do certain tasks that fall within their specific job description. When, as workers, we are forced away from performing our value-added functions, we begin to feel overworked. As a production worker, your job would be to build a product or keep a machine running making parts. If line workers must stop to search for tools, retrieve material, rework mistakes, or any other waste that pulls them from the process, you can bet they are working hard. Working smart is always the model to shoot for, and working smart is being in a process that allows you to perform value-added work.

If 50% of the day is spent in waste and the line is expected to complete a desired output for the day, boy, that will be one overworked group. I hear all types of comments when working with employees who have been stuck in inefficient processes. "We do what we have to do to get the job done." "Make it happen." These types of comments simply tell me that they have to deal with inefficient processes and, typically, they have been stuck in them for some time.

Speed!

Speed gets you nowhere. Batch-processed environments produce huge lots of product before changing to another type. The idea is to keep the machines running making work-in-process (WIP) or finished goods. Plus, the faster the speed, the more product and the better utilization of the machine, right? Well, there is some minor truth to this, but there are also a few problems. First of all, it makes sense to keep the machine producing products so you can get a return on your investment. The problem with speed in any process is that it may hinder quality and jeopardize the safety of the workers. Often companies offer financial incentives for output and this creates even more problems. A door manufacturer in Oregon was caught up with output and they ran their production lines quite fast. People had to be busy doing something because management had the misconception of "staying busy." As long as the person was doing something, value added or not, they felt labor dollars were being maximized. The equipment was set at high speeds and the operator worked hard to keep up. Everyone had the mentality of speed and staying busy. Plus, there were some great incentives for "punching" out a lot of product.

This approach created a high quantity of WIP and finished goods, depending on the process. The company also spent a lot of labor hours in reworking doors due to dimensional errors and visual blemishes on the wood. They actually had a rework team that worked every weekend fixing doors that were produced incorrectly during the normal work week. Incentives continued to be offered for output. As the months passed, inventory continued to grow. Then the recession of 2008–2009 came along and they were stuck with a lot of finished goods inventory taking up floor space. Good example of staying busy. My point in this example is that the process was set up to overproduce, make mistakes, and create an environment that encourages speed.

Another example comes from an organization that produces a wide variety of residential cleaning products. The majority of the work performed in the facility was filling plastic bottles of product, placing a label on the bottle, and capping it. The value-added portion of the cycle was quite short. This value-added time I am referring to is the actual work performed to change the product: Fill—Label—Cap—DONE. However, this simple process was made very complicated and lengthy. To keep people working hard, the production supervisor would have one or two operators prelabel the bottles weeks before they needed to be filled. They operated an automated labeler in which the operators placed the bottle into the machine, pushed a button, and the machine did the rest. The supervisors would just keep giving them bottles to label and sometimes the workers would prelabel thousands of bottles. Often competition would arise if there was a second operator, and they would see who could work the hardest and fastest. Anyone who could not work at this extreme pace was considered a lower-level worker. The supervisor kept them busy by shoving bottles in front of them, telling them to work faster. By the end of the day, there were dozens of boxes full of prelabeled bottles or, in this case, "partially" complete product that could not be sold. In the operators' eyes, they worked hard and, quite honestly, they did, but for what? To keep a supervisor happy?

The boxes were then placed in the warehouse and proceeded to sit there for weeks until it was time to fill them with product. Many occasions arose when the bottle fillers pulled out the prelabeled bottles and the labels were not straight, torn, or the bottles had an incorrect label. Keeping workers busy is not necessarily a good thing, and it is a sign of a dysfunctional process.

LIVING WITH WASTE

I used to work with an engineer who, like myself, spent the majority of his professional career in manufacturing and was a serious practitioner of Lean. We worked together in an assembly plant in North Carolina in which we helped turn around the company, top to bottom, using Lean in the office, production, R/D, maintenance, and every department. After about a year, we were having lunch, talking about the next step for the organization, and he said, "There are three things you cannot avoid in life: taxes, death, and waste." He was referring to the amount of inefficiency in the company that we still had to reduce or remove. He was referring to how people in the company simply live with waste and what actions become the side effects of the company. To further illustrate this, I will use the first seven wastes as my guideline, excluding human potential for now.

Living with Overproduction

Organizations that are living with overproduction have an array of problems. Overproduction is by far the most common of the eight wastes and creates all the other seven. Manufacturers that are heavily equipment based or have mostly automated processes often are living with overproduction due to downtime. Downtime, no matter what the reason, adds no value to the product, and your customers are not willing to pay for it. Downtime can be from setup, quality problems, and machine issues.

Setup

Setups are absolutely necessary in any equipment-driven process, but the actions themselves are nonvalue added. When companies do not challenge current setup and changeover techniques, they create an environment to overproduce, and many people and departments have to live with it. There are very few companies that market and produce just one product. Options and variations to product offerings are plenty. If you offer a multitude of products and operate a machine-based process, changeover is critical to meeting customers' orders, reducing inventory, and decreasing lead times.

When stuck with long setup times, manufacturers avoid changeover because they fear the downtime. Once the machine is set up, the company will tend to build as much as it can until the next changeover is necessary. To avoid changeover, companies produce large lots of product, WIP and finished goods, to compensate.

Often, they will look farther into the schedule and see if the same product is coming up and simply build that order as well. Meantime, the machine is tied up making product that is not needed, at least in that quantity, and unnecessary material is being consumed. But what happens if there is a quality error? The entire lot may be defective. By avoiding improving setup times and developing quick changeover techniques, the machine operators will be forced to live with overproduction.

This trickles over to the purchasing side of the business as raw material is used sooner than needed, and material must be purchased in larger quantities. This spends working capital for other things needed in the organization. As the overproduction continues, buyers will live with the problem and be forced to purchase "unneeded inventory."

Quality Problems

It may surprise you but outside of surprise defects and rejects, I have witnessed companies that overproduce to compensate for planned defects. By utilizing bad or outdated equipment, there is an increased risk of making defects. Rather than fix the problem or make the proper modifications, additional product is made to ensure the correct number of good parts. If you have this scenario, then the company will have to live with overproduction.

This is the same case in a manual production process. A manufacturer of composite material for the boating industry created an environment where 30% of the parts required rework and they purposely manufactured 30% extra to compensate. Tooling design was the problem; incorrect contours were made on the individual parts. The way the process worked was that a work cell created individual flat and contoured parts. Operators would take the squares of material and cut them to specific square dimensions. They also would take some of these flat parts and place them over a contoured fixture and reform them. Both the flat and contoured parts would go to an assembly area to make the major components of this special type of boat.

The company knew of the 30% rework problem on contoured parts. However, they avoided fixing or even reducing the occurrence. The side effect was to not only make 30% more to compensate, but the contoured parts needed specially designed storage racks to maintain their form. Living with this overproduction forced the company to physically alter the layout and floor space used to allow for it.

Machines Issues

Outside of a machine creating defects and rejects, when organizations do not maintain these vital assets, machines can shut off due to mechanical problems. The backward thinking on this is that some will simply produce more when the machine is "in control" when there are no immediate problems. Knowing a machine breakdown is coming, this overproduction in their minds is creating enough inventory to balance out any upcoming loss of volume. This example is not much different from buffering for quality problems, but I am sure you see my point. Like all waste in a process, we just live with it.

Living with Motion

Oh, this is a fun one. I say this because living with motion is extremely common and often easily fixed. Bordering on ridiculous, operators constantly live with motion. In a production process, motion comes in a variety of ways:

- Sifting through piles of tools
- Leaving a workstation for parts
- Looking through a box of parts
- Walking to and from different departments
- Looking for items in an unorganized work area
- Looking for tools that are simply not available
- Searching for a supervisor

The list can go on and on and represents a series of necessities simply not readily available, and I mean within five feet of where one works. You have to apply common sense to this one. Would you prefer that everything a value-added worker needs is within arm's reach, or have that worker constantly stop producing product to perform the above list of motions? Lack of organization in the visual workplace creates a ton of motion. Living with it simply creates a burden on the operator. Thinking back to Chapter 3, excessive motion makes workers feel worn out. Loss of focus on quality and decreased cycle times can make the operators feel as if they are "working hard to really accomplish nothing." Thus, left unfixed, the workers live with it and find ways to cope.

I have a clear memory of a technician who worked for an auto body shop and lived with motion beyond anything I had ever seen. This individual worked in the dismantle area, or sometimes called the blueprint process. The blueprint process is the first step in the repair flow in which highly trained technicians evaluate the incoming damage and compare it to the estimate that was created in the front office. Part of the job is to check what parts may be missing on the estimate, and there is often hardware, small plugs, clamps, inserts, and many other items that are not identified. This first step is considered a double-check for a very important reason—because once it is signed off by the technician and the automobile is dismantled, the repairs begin. Therefore, catching potential mistakes at this point is critical. Due to the inefficiencies in the front office, mistakes in the estimate, the distance to the blueprint work area, and the lack of organization in the work area, this man and his fellow technicians walked hundreds of miles a month. And, well, they lived with it. In essence, the technician walks and retrieves the car, walks to the office to see if the estimate is ready, and walks back with no estimate. The office calls for him, so he heads up front, only to find out they had a question about this particular car, so he heads back to the shop. He begins sifting through tools and looking for items missing in his work area to prepare for the estimate arrival. He walks back up to the office, finally grabs the estimate, and walks back. He then proceeds to review the estimate and compare it with the damage. He starts to make the changes and walks back to the office, drops off the estimate for revisions, and proceeds back to the work area. He begins working on a different car and then gets called back up front. He retrieves the estimate a second time and walks back to the blueprint area—and on and on and on. Do I need to continue? Imagine living in this environment and watching as no improvements get done.

In administrative functions, motion is a reality of life when electronic information on servers and shared network drives are unorganized.

Think about your public drives. Ever feel like you spend more time looking for the information than actually using it? As public e-folders become more and more dysfunctional, more time is spent away from their use. Rather than organize and agree on what needs to be sorted out, people in offices will start to create more folders from scratch and begin them. This takes information out of standardized folders and it becomes something more customized to a person's liking. Do you want your administrative professionals working on things like marketing plans, new product development, cutting checks, creating invoices, and creating estimates? Or, living in this type of motion?

Living with Transportation

As a reminder, transportation is different from motion because it is the movement of product, folders, parts, information, for example. Often it requires a device to transport, such as a cart, forklift, pallet jack, or dolly. Companies spend a lot of time, money, and labor simply moving product around the plant. Recently, my team completed a full analysis of a product family from raw material buying to delivery of final products to the customer. One important evaluation is the identification of movement of material throughout the process. At the end of week, it was identified that, on average, material travelled about 17 miles each day. In a given year, this equates to around 4,200 miles (or like driving from Atlanta, GA to Anchorage, AK). This is a result of what is called process-based manufacturing and it can create issues more than just transportation.

Process-based manufacturing is the opposite of Lean work cells. In a process-based environment, work is performed in certain areas based on the type of discipline and then moved to another area for the next type of discipline. For instance, there may be a welding department, a grinding department, an assembly department, a paint department, a packaging department, and so on. Material is processed, then generally stored, moved, stored, and then finally processed again. Moving away from this scenario is critical to becoming a Lean organization, but, more important, for significantly reducing lead times, improving productivity, and reducing WIP levels.

Often, a process-based manufacturing environment is created over time as the company grows. Employees and the company as a whole are forced to live with transportation. From an organizational approach and as the business grows, you are forced to buy all the transportation devices required. Forklifts, for example, are not an inexpensive asset. Outside the initial cost, there is maintenance cost, driver certifications, ongoing operating cost, floor space to store them; and, as the company grows, so does the need for more of these costs. Often these transportation devices are not readily available, so operators may leave the work area to find them or improvise, forcing them into potential ergonomic problems. I have witnessed when forklifts, pallet jacks, or other devices were not available; but rather than wait (which is equally bad), the product was moved manually. When asked why they pushed or pulled the material around the plant, the answer was that they get tired of looking or waiting for a forklift. Well, the missing forklift is the side effect of transportation. The irony of this problem is that now the company is forced to hold more inventory staging points throughout the facility. When processes are separated by long distances, incoming and outgoing inventory points are needed. There is no balance or flow to process-based production because there is no visibility of what is coming, so inventory collection is needed. As forklift drivers drive into an area, he/she must have a place to drop off the material.

Living with Overprocessing

As mentioned in Chapter 3, overprocessing truly is the waste of redundancy. We see it in fabrication departments, administrative functions, processes requiring skill, and again in automated processes.

Administrative Functions

Office processes and paperwork-driven functions are often riddled with overprocessing, and living with it is a daily challenge. Overprocessing occurs here when redundant effort and extra steps happen causing the "service product" to slow down through the process. Service products are generally nontangible, such as information flow, but still can have tangible elements. For instance, a repair estimate in an auto body shop is a piece of paper that contains all the vital information about fixing the damaged car. For example, work orders, check and invoice creation, developing a marketing plan, and new product development all contain what we call service products. There is a process for everything, and these administrative processes take time, involve multiple steps and multiple departments, and pass through numerous people. Administrative processes simply contain redundancy.

The redundancy comes in the form of extra checks and verifications: printing multiple copies of the actual service product, waiting on several approvals, and plenty of other excessive steps. In the beginning, processing the service product may have been quite simple and moved through in a short amount of time. What often happens is that a minor mistake is made along the way that either creates problems for the end customers or some other department internally. Now, people become scared and create a ridiculous number of checks and balances to ensure it never happens again. Multiple sign-offs are now needed, and employees begin saving e-mails with vital information to protect themselves. All of a sudden, what used to take a short amount of time now takes much longer. Nothing really is fixed and it becomes part of the process. Needless to say, new employees are now trained to deal with waste.

Extra Steps and Redundant Effort

Waste is like a virus, simply growing and multiplying over time as processes become less and less efficient. Companies with poor quality systems tend to add on layers of extra checks and verifications to ensure defects or other quality-related issues do not make it downstream. Often, unnecessary checks are created because of a one-time mistake that really upset a customer. Not to imply that reacting to legitimate customer complaints is not good business practice, but what truly happens is that this additional waste is in place to ensure it does not happen again. Sure, placing systems in the manufacturing process to ensure, at a minimum, nothing of poor quality gets into the hands of the customer is a good thing. Stopping or relieving the symptom is good, but has anything truly been fixed at the process level? Root cause analysis of the problem will truly eliminate or significantly reduce the occurrence. From a Lean perspective, the process should be evaluated to find the cause of the waste, and process improvements performed to eradicate it.

Without continuous improvement, two wastes now exist: (1) the cause of the problem (defect) and (2) the extra check (overprocessing). Operators must now live with uncertainty and be required to perform added verifications of their work or a machine. Additional paperwork may be needed, and inspectors may need to perform more inspections. The quality assurance

(QA) department now may have added responsibilities from an oversight perspective.

Inefficient Machines

Many organizations overlook the importance of machine optimization, usually because they confuse optimization with utilization. Machine optimization is about increasing the longevity of the company's assets that perform value-added work. When machines are not maintained properly due to poor preventative maintenance programs and inadequate operator training, the machine will simply work harder to get the job done, harder in terms of lower quality and longer cycle times. Here is a great analogy.

Most of you reading this book have mowed a lawn with a lawn mower. Living in the Pacific Northwest, I don't have the luxury of a consistently dry lawn. Mowing a lawn in a dry environment versus a damp environment has vastly different impacts on the mower. The mowing season starts around May and concludes around early October (here in Washington). As the season progresses through a mix of wet/dry grass, the mower goes through general wear and tear. Grass accumulates under the mower and on the blade. Bolts on the handles become loose and worn out. The blade becomes dull, and the wheels begin to make a squeaky sound. As the grass continues to accumulate, it becomes harder to make its way through the exhaust and leave the mower. Often it makes a "puking" sound as it stalls and then relieves itself with a large clump of grass. As more weeks pass, the mower will stop running or slow down to a very low rpm.

If I don't maintain this mower, clean out the grass, tighten loose bolts, and oil and grease squeaky areas, it is going to work harder to get the job done. It is overprocessing. By the end of the season, it probably takes me 20 to 30% longer to the mow the lawn when compared to the beginning. Eventually the mower will wear out sooner than its intended life span. Silly example, but I think you see my point.

Machines and equipment areas are vital assets to any manufacturing company and should be viewed that way. We witness time and time again the lack of care of these assets—either from no or very little preventative maintenance programs to lack of accountability and visibility that the actual preventative maintenance is getting performed.

Inability to Identify a Completion Point

Fabrication

Overprocessing occurs in fabrication departments when jobs such as sanding, cleaning, deburring, for example, are overperformed. Another way to look at it is processes that have unclear endpoints, and meeting tolerances is in the eyes of the actual worker. Hence, you can argue that living with overprocessing may not be reduced because certain jobs just require an expert eye to decide. There even may be a range of that expertise in the process, but as long decisions fall within that range, everything is fine. No argument. However, sanding, cleaning, and deburring, as my examples, do require tools and supplies to perform the work. Ever tried sanding a piece of wood manually with a piece of sandpaper worn out and torn? Tough, right? Maybe you are thinking that by using an electric sander, you can reduce the overprocessing. It's possible, but now you have to maintain the sander because it also can get worn out and take longer to perform the job.

My point is that you may have very skilled people in these positions that require that expertise; but if they are working with inferior tools and supplies, they will live with the overprocessing.

This element of overprocessing is often overlooked. We discussed early in the book that everything in a business is about reducing the gap between the inputs and the outputs of any process. Every process has a clear start and endpoint. As you place parts into a car going down the assembly line, you can witness its completion because these components snap together or are installed with some form of tool. Auto body repair shops have similar processes as parts are reinstalled onto a now-repaired car. Office processes also have definable endpoints. As information is placed onto a work order, such as customer information, job numbers, due dates, etc., the office worker can literally see it coming to completion. However, there are many situations when identifying the endpoint, at least consistently, is difficult.

Many operations rely on the expertise of the worker because finding this completion point really is in their hands. How do you know something is cleaned, sanded, polished? Just a few examples, but if you had four people performing these tasks together under the same conditions, they will end at a different time. I like to use these three examples when discussing how people live with overprocessing. The reason is that each one requires some type of material, tools, and supplies that can have a profound difference on cycle time.

A lot of our clients have operations in which sanding in some form is common. Poor investment in the right tools and supplies will extend lead times and create frustration for the workers and they will improvise as needed. One organization in Arizona had processes that required the operators to sand down different types of fiberglass-reinforced material into various contours and shapes. The company had not replaced or maintained the tools in a long time. On top of that, standardized tools were never identified as each operator was using a different style of sander and even sanding disks. Each phase of the contouring process truly required different grits and styles of sandpaper. Over time, operators began to bring in their own tools of preference, use whatever sanding disks were available—often stealing from the maintenance department because maintenance was known for having the best disk type available. Sometimes, operators would not even use the sander because they felt they could do the work better by hand. This would, of course, extend the cycle time, but quality was of the utmost importance.

Operators often complained to management about the problem and there were random localized improvements, but the overall problem was still there. Variation already exists in operations such as sanding, but to cope or live with this overprocessing, more variation was added to the whole process. My point here is that people will improvise their work environment to deal with these issues, and this reduces their confidence in the company's ability to invest in the process.

My final example of living with overprocessing is when companies apply random and often very long cycle times to operations such as ovens, curing, and drying processes. Again, a different approach when looking at waste, but when looking at multiple industries, I see a common theme. In operations where product must be placed into an oven for drying or curing, why is it that over many different industries, everything takes 24 hours to cure? It's quite comical.

We see time and time again, these long cycle times for these types of processes. I realize that in a lot of industries such as aerospace, auto body repair (painting), and fiberglass, cure is essential to overall quality and durability, but why these random times? Is it 24 hours, or is it 18 hours and 45 minutes, or whatever? Engineering departments are often the ones responsible for establishing these cure times and, to play it safe, which makes sense, they select longer than needed measurements. Curing and drying, for example, are often debated in the Lean world about whether they are value added or nonvalue added. I am not going to get into that right now, but to make all parties stop debating and get a point across, they are value added to a point and then it becomes overprocessing.

Living with this type of overprocessing creates a much larger issue than the sanding example. When production processes such as ovens have these 24-hour cure times, organizations use this constraint to control flow. The processes before the oven will overproduce and create larger lots of product to be staged in front of the oven. This WIP quantity is also established not from the 24-hour cure time, but from how much can fit in the oven. With WIP, you create wait and queue times that add to the overall lead time throughout the entire facility. The manufacture may even make adjustments to the shifts people work and the sequence of work because of this 24-hour cure time. An array of larger adjustments are performed, which is my point. Keep in the mind that you do want to keep your constraint full of product so that it does not run out of work to do, but I then challenge the need to apply these large cycle times.

Living with Defects

Companies all over put up with quality problems in regard to defects, scrap, rework, and general mistakes throughout the entire value stream from suppliers to customers. Like any waste, there will always be some level of quality problems in any process: manufacturing and administrative. It's the tolerance of extreme quality issues and something companies and people just deal with. I could probably write an entire book called *Living with Defects,* but I will keep this short and to the point.

Poor quality work can come at any point in the process and reducing this waste will be ongoing. Even when you get to a point where you have minimized defects and errors to a small percent, you should continue to drill it out of the process.

Some time ago, I was part of a team of Lean practitioners who were hired to come to a company that had some serious internal quality problems. The organization was located in San Antonio, TX, where they produced fiberglass reinforced piping for the oil and natural gas industries. Much of the process was manual, but when dealing with resin and catalyst to bond fiberglass, there were many factors that contributed to the quality of the final product. Room temperature, air quality, proper mixing of resin and catalyst, operator skill, and moisture all contributed. Conditions did not need to be perfect, but the right mix was needed. You could imagine the amount of overprocessing in this environment as well.

The first part of the operation was to produce a certain diameter of pipe that eventually would be joined up with flanges and elbows to make a final assembly that is installed in refineries and power plants. The flanges and elbows were produced in what is called a fitting department, and then all components including the pipe would be transported to the assembly area. Problems existed in the flange and elbow department.

The first process was to mix resin and catalyst and then to place the mix onto pieces of fiberglass. The now-wet fiberglass would go onto a mold to create the form of the elbow and flange. Several pieces of material were placed on it until a certain thickness was achieved. Often, air bubbles and other visual defects were identified after the curing of the parts. Often, and being conservative, 65% of all elbows and flanges were considered defective. Of that 65%, about 70% were reworkable and the others had to be redone. Air quality was poor in this department as dust and other particulates from grinding and sanding would contaminate the molds. There were no standard guidelines for mixing resin and catalyst. New workers relied on the experienced employees to show them how to mix components. However, even then, there was no real guideline.

The room was next to the receiving doors so because deliveries were made all day, cool air from outside would come into the room and change the room temperature, blowing dust all over the room. The molds were old and needed replacement, and the tools that were used to roll down the fiberglass were old and required the operator to spend a lot of time working on each component. Rather than incorporate continuous improvement and fix the problems, here is how the workers lived with it.

First, extra molds were purchased over time so workers could overproduce and make more flanges and elbows than needed to compensate for lost components after failing the curing process. A percentage of what they called "excess" was added to the work orders from customer service based on an estimated percentage loss due to scrap. Rework stations were built into the room to repair defective flanges and elbows. Rework stations also were placed in assembly so, as assembly workers would check the components before attaching them to the pipe, they could rework them rather than having to send them back to the fitting room.

The company had to purchase racking and build wooden shelves to store the added fittings. Depending on the customer requirements, these overproduced fittings were not useable and would often sit for months before being thrown out. Due to a lack of organization, assembly workers would sometimes grab the wrong flange or elbow created in the "excess," and then the final inspector would catch the error and major rework was needed. The company just kept adding more and more buffers, and workers improvised throughout the entire process.

The typical lead time from when the material was retrieved until final assembly was somewhere around 6 to 10 hours. Fifty-five percent of the overall cycle time was the result of rework. Nearly half the time was spent repairing components. Fortunately, the customer never received unusable product so, of course, the company felt its efforts were worth it. They were proud that nothing bad got to the customer, and they should be. But, without implementing some improvements, they were living with defects that were costing them a lot of money.

Living with Waiting

Improvising in the face of waiting is not generally as bad as the other waste. Waiting often occurs when the required items are not readily available to use. Lack of timely information, tools, procedures, material, people, and product can make workers wait. Generally, it is not as obvious because what happens is they will find something else to do. On the surface, this may appear to be a good thing. Having multitaskers and cross-trained employees is extremely valuable. It becomes more of a problem because rather than waiting on what is needed at the time, they will begin to overproduce by performing work that is just not necessary. This is not the solution to waiting; it's a side effect of it.

Employees, especially in production-type environments, do not want to convey the appearance of having nothing to do. To deal with the waiting, they start to "look busy." Looking busy can be a very inefficient thing. For example, let's say a machine operator is waiting to set up and run a quantity of parts on a computer numerical control (CNC) machine. All the required paperwork, fixtures, and tools are ready to be used, but they are lacking the raw material. They wait, walk around asking questions, and at some point give up and proceed to set up for another run of parts. These parts may be needed at a later time, but at the moment, not necessary. Now, the machine operator needs to store these items somewhere when complete. Two wastes were just created: overproduction and inventory.

A potential third waste that can be created is defects. Let's hope that the lot of parts, which were overproduced, has no quality issues. Waste just creates waste, and this is the most common side effect of dealing with waiting. The worker could start walking around as well to find something to do or to ask a question about the material, but you see my point. Often it is less costly and smarter for the worker to stop and find out what is wrong rather than stay busy. Staying busy can become a very costly practice as the activity consumes inventory and labor hours at the same time. The long-term solution is not to embrace staying busy, but to improve the process to ensure that value-added workers have everything they need to perform their work. More to come on that.

Living with Inventory

Question: Do you build product or store product? Your inventory may be defining what your business truly is. As a producer of tangible products, you have to get control of your buying habits, supply chain, production processes that create WIP, and finished good levels. Living with inventory is a way of life for many organizations. How they deal with it usually creates the framework and foundation in which the company operates. As discussed in earlier chapters, there often is a gap between what companies buy and what they consume. When you have a large gap, it means you are tying up a lot of money in excessive raw material and are depleting that inventory at a slow rate.

WIP accumulates in between production processes due to long travel distances in between those processes, imbalances in work content from one process to the next, lack of needed items, poorly trained operators, and poor sequencing of work. There are plenty of other reasons, but these are the most common. Finished goods pile up when companies build for the just-in-case scenarios and because their production processes are not set up in relation to the pace that customers are purchasing product. Here is how companies live with the waste of inventory.

Facility Space

Well, you are now going to need space to store the raw material as it comes in, longer production processes to accommodate the WIP, and even more space to store the finish goods. What also happens is that the company is forced to physically lay out the production area "around" the inventory. We see a backward approach to facility layout when there is a lot of inventory. The preferred approach to facility layout is to establish the best layout for the production process to promote smooth flow, reduced cycle times, visibility, and reduced travel distance. Then, all required material and support functions are placed around the production process. The production process is the value-added activity, and all essential functions and necessities are placed in a way to support it. We see a lot of companies design the physical layout to accommodate the space taken up by inventory and this is backward thinking. However, when you have a lot of inventory, at all three levels, you become forced to deal with it in this manner.

Production Process

Dealing with WIP is a pain in the butt…period. It is the source of so many problems and creates huge delays in lead times, increases travel distance, and ties up money in labor and material. Production workers deal with this by requesting and placing added work surfaces, tables, racks, and other storage devices to store the WIP being created. Makes sense because they need to quickly move on to the next product. I have witnessed so many makeshift tables over the years to accommodate WIP. Pieces of plywood and on top of garbage cans are the most common, either to store WIP or as a makeshift work surface because the original work surface now is full of WIP. People get very creative with this type of improvising and, as a leader, if this is the case in your plant, you are creating a culture of improvisers. Anytime you have to live with waste, you see this improvising happening.

I can recall a situation in North Carolina when I was helping an assembler of coin dispensers implement Lean plant-wide. During some initial analysis, my team observed operators working in a makeshift station. Due to the excessive amount of WIP being created in a subassembly work cell, space was becoming limited. The operator had taken a control panel subassembly and placed it on top of two small pieces of 2 × 4 wood to construct a table. They were using this table to build other subassemblies intended for final assembly. I am sure that at the moment of conception of the idea, it made sense. They were being pushed to build, build, build, and the available work area was shrinking by the hour. This was the way they dealt with the WIP inventory.

Living with waste is a sad thing, and it is critical from a psychological perspective to incorporate Lean to reduce this waste. You do not want to create a culture of improvisers. Improvising should be the exception to the rule, not the norm.

5

Making Change Happen with 5S

The successful implementation of Lean into the overall business will help change the way your culture thinks and perceives the company. This chapter is dedicated to providing one powerful Lean implementation tool that will transform your business and to show how this will change a lot of the dysfunction I described in the first four chapters.

Within the Lean philosophy, there are numerous tools and techniques for reducing waste, and each one has a place and purpose. Many organizations today starting Lean struggle to find legitimate starting points with Lean and often the journey fails quite fast. Don't get caught up in the giant picture of the overall business model of Lean; start small and work your way through it.

This chapter is solely dedicated to making change happen with 5S (sort, set in order, shine, standardize, sustain). It deserves its own focus as the next chapter will dive into many other Lean implementation tools beyond 5S.

5S AND VISUAL CONTROL

Let's start with the most important implementation tool of them all. To give you the magnitude and potential impact 5S has on a company, two years ago I co-authored a 175-page book solely on this practice. It is by far the most common implementation projects I lead. It is the starting point for all other Lean initiatives, and it is always changing in a company.

5S can be applied in virtually any environment: manufacturing, assembly, R/D, maintenance, shipping/receiving, administrative functions, auto body repair, auto repair, warehousing; the list can go on and on. 5S implementations generally follow a common pattern, but, like all Lean methodologies, it must be tailored to fit the company and specific process.

5S is an aggressive visual organization philosophy that is implemented to create a highly visible, pristine work environment. Visual control is the second layer of the visual workplace that creates a level of accountability in the process. Many of you reading this book are quite familiar and even experienced in 5S. Although I will describe the concept in detail, I also will discuss how it will improve the thinking of your culture.

One of the reasons 5S is a great start to Lean is how it tests the culture of your company. Many Lean methodologies require significant data analysis to design Lean processes. Establishing volume requirements, manpower needs, equipment utilization, when to consolidate processes, inventory reduction plans, standard operating procedures, and many more, is what I call higher-profile implementations. 5S is the creation of the visual workplace, and you can see early on how everyone will adapt. Resistance and buy-in will come in many forms, and 5S will set the tone for the level of each. I describe this as I go along.

5S can have a profound impact on productivity, floor space use, product and people travel distance, inventory, and quality. 5S is also the tangible element of Lean that often is proof that when Lean is practiced in a company, it takes workplace organization to a whole new level. In a 5S-compliant environment, everything deemed necessary has a home with clear visual markings to its locations. Here are the 5S:

- Sort
- Set in order
- Shine
- Standardize
- Sustain

Sort

Sorting is the first step not only in 5S but in your overall Lean journey, and you will find as you move along every year that you could become possessed with sorting. Sorting is the act of discarding all unnecessary items from each work area and eventually the overall business. As time goes on, "stuff" simply piles up, people hide and horde things, and the workplace becomes generally inefficient due to disorganization. In preparation for the visual workplace, the act of sorting is done as a detox to the area.

The first cultural battle you are going to find that you must win in the beginning is the need to hold on to everything. Maintenance, offices,

FIGURE 5.1
Example of a sorting project.

and R/D departments are generally the worst offenders of this, and you will find having them commit to sorting may be a challenge. Major sorting is really the key here to break this trend and challenge everything in the work area. Never move on to the next "S" without a thorough sort. A workplace should only have the bare essentials. Some departments, e.g., maintenance, may require extra parts and tools, depending on the number of technicians and lead time from suppliers on equipment parts, but production areas should be cleared out. Figure 5.1 and Figure 5.2 show the results of sorting in a real workplace.

Two types of items should be removed. The first is garbage. You would be shocked at what has been sitting round and not touched for years. No point in even having a battle over this; remove and toss it. The second is items that are not needed now. Probably the most common items removed are things really not necessary anymore, but this may need to be evaluated by management.

Do not let the cost of something stop you from removing it. If a brake press has been sitting around for three years in a fabrication department due to a newer model, remove it. No one is saying that you must throw it in the garbage, but make a management decision on what to do with it. An organized approach to sorting high-dollar or potentially "useful" items is through a red tag event. Simply place red tags on the items being removed and place them in a temporary holding area for further analysis. This is a

FIGURE 5.2
Another example of a sorting project.

visual statement saying that an item is no longer needed in the work area; it was removed during a 5S project, and a final decision should be made on its fate. Figure 5.2 shows this system in use.

I also recommend that during your sorting phase you even remove necessary items temporarily in preparation for setting up the work area in a whole new layout to better improve flow, communication, visibility, and to reduce cycle time. A common method of sorting is to just remove unnecessary items. I like to see a complete clearing of the area and then it is pieced back together in a way that reduces waste even further.

When you take this approach, your organization is making a clear statement that the old way of working is gone forever. Nothing opens the eyes of employees when the work area is there at one point, and then it is gone the next. It really prepares the implementation team for a new layout, and it allows the company to start fresh.

Sorting can be very difficult for people. Like any "detoxing," letting go of the things that have been held onto for years may be challenging for some. People are hoarders by nature. Look at the ever-growing self-storage industry. Human beings spend a lot of time and money storing stuff they don't use. A business is no different. Look across your office, body shop, production floor, or whatever you do—stuff is everywhere, and this stuff simply feels good. As you begin sorting, you will have to deal with those who do not want to part with these items. Thus, your first Lean test begins.

Some of your people will be quite frightened by this approach, and I have often see nervousness in the eyes of the team. On the other hand, a good portion of your culture will feel quite invigorated by this approach and embrace it with open arms. "Set in order" can truly now begin.

Set in Order

With a clean sheet of paper in front of the 5S team, you can now set up the work area. There are a couple approaches to this next step. Lean is very data driven; however, for the purposes of 5S, often data are not needed to set up the work area, especially if it is in a nonproduction-related environment. When implementing Lean into an assembly line or other form of production, many other Lean tools are applicable, not just 5S. Even one of our existing clients told me one time, "If we just focused on 5S and 5S only, I would see remarkable measurable improvements to the company." He was right.

You can begin set in order with little to no data. Although a general discussion of workflow, point-of-use presentation, and worker motion is the best preparation for set in order, I have witnessed 5S projects that did nothing of the kind and were highly successful.

Set in order is the act of visually organizing the work area through the use of floor tape, paint, labels, designations, right sizing, and signage. So, when I say "everything has a home," I mean it.

The set in order phase starts from the floor and moves upward so that general layout and the identification of essentials, such as workbenches, trash cans, pallets, inventory, and dollies, are addressed first. With the area completely cleared out, or at least the best you can, put together a layout that reduces transportation, motion, and overproduction. Establish where people will stand and ensure that necessary items that do go on the floor are within five feet of the worker.

The key to the new layout is designing it so it is physically different from the way it was. Challenge yourself to smaller work surfaces, eliminating cabinets and shelves, change product flow, or whatever makes sense for waste reduction. From a psychological perspective, when you change the physical surroundings of the workplace, it is nearly impossible to go back to the old way of working. The big giant table that allowed the worker to overproduce is now gone. All the bins that used to accumulate excessive shop supplies are gone. Horizontal surfaces have been reduced so junk cannot accumulate.

As humans, we act and react to the geography that is presented to us and when your 5S teams change it dramatically (for the better), you automatically change thinking. If the worker tries to work in the old-fashioned way (i.e., producing excessive WIP, hoarding supplies, piling tools on the workbench, etc.), it will be easily visible and the process won't work.

After the team agrees on the new layout, then begin floor taping all the items on the floor to create visible home locations. You may need to paint lines instead of floor tape, depending on the floor's condition. Also, when finalizing the locations, make sure to create addresses and designations for everything. For instance, designate a letter and number to the workbenches and shelves. Figure 5.3 illustrates this concept and Figure 5.4 is a great photo of the concept in place.

As you progress upward, create shadow boards that will be used for visually storing necessary items in or near the work area. We call this "going vertical" and it works.

Going vertical challenges the need for horizontal surfaces that simply take up floor space and are breeding grounds for stuff. Figure 5.5 is an example of a 5S-compliant shadow board. You can see that tools and supplies are placed on the board. The shadow board also contains a designation so if the tool is found in another department, the marking on the tool can direct it back to its home location.

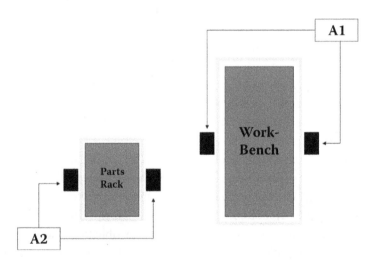

FIGURE 5.3
Graph illustrating the designation of letters and numbers of workstations and shelves.

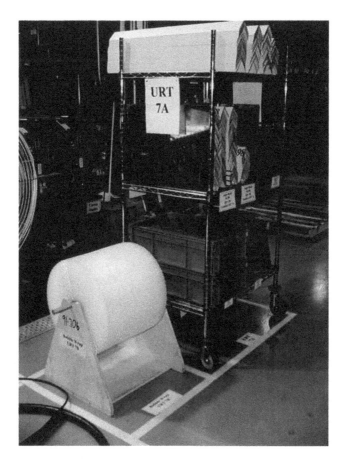

FIGURE 5.4
Photo of final home location.

How to Make a Shadow Board

1. Paint the board: Any color will work. We have seen red, yellow, black, blue, green, and orange shadow boards. In fact, colored "peg board" boards are available for purchase; therefore, you can pay a little more up front and save on the painting step. My personal favorite is the industrial gray diamond-plate look.

2. Create the tool layout: Once the tool boards are dry, place them on the floor, a table top, or another flat surface and lay out all the appropriate tools and supplies on the board. These boards are excellent for holding items such as tools, materials, tape, scissors, calculators, clip boards, and other shop necessities. Your supply box should contain peg hooks, double-back tape, Velcro tape, and other items designed

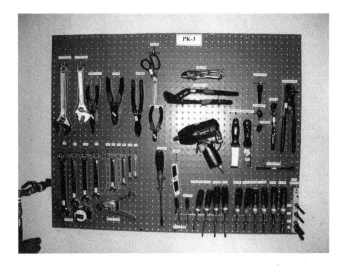

FIGURE 5.5
Photo of a 5S-compliant shadow board.

to hold things vertically. Take these things into consideration as you make decisions about where items should be placed on the board. At this point, you are designing your board layout and identifying how much board surface will be needed to hold the workstation's supplies.

3. Cut the board: Now remove the tools from the board, then cut a square or rectangular piece of board that is sized to accommodate the footprint of the workstation tools, parts, and materials it is expected to hold, while also ensuring it will fit on your allocated wall space.

4. Install the tool board: Construct a wood frame for your board using 1″ × 1″ studs. The board needs to float at least 1 inch from the wall in order for the peg hooks to have clearance and lock into place. Walls are the logical choice for tool boards, but actually these boards can be placed anywhere. We have seen them installed on the side of workbenches and cabinets, on the back of HVAC units, and mobile on casters (wheels). Just be sure that your board is visible, accessible at point of use, and does not impede or disrupt the flow of product.

5. Hang the items: Have your team hang the tools and supplies on the tool board. Be sure to leave room for labels in between the items and space to draw an outline or shadow of each item. This process is relatively slow, but it is the key to this part of the 5S implementation.

a. Tool size: Try to locate heavier hand tools lower on the board, but make sure they do not get placed on the floor. Not only do we need to limit the effort needed to move heavier tools to and from the work area, but we don't want a heavy item high on the tool board for safety sake because the item can fall down on the person retrieving the tool. On the flip side, lighter tools can go up higher on the board, yet stay well within ergonomic reach. The most frequently used tools are normally placed nearest the point of use.

b. Tool sequencing: Placing tools in a sequence, mirroring the order in which they are used, may also make sense for your application. In other words, when the first step in a documented procedure calls for screwdriver SD123, place screwdriver SD123 first in your tool sequence on the tool board. Likewise, if diagonal cutter DC1A is required second in the process step, have it strategically placed in the second position on the tool board, and continue until the entire process is complete.

c. Include everything: Find a home for every needed item on the tool board(s), including, for example, the pen used to fill out a quality form or the work cell calculator, which is also in need of a home, with each item being located and labeled on the tool board until every one is accounted for.

Assuming the 5S sorting activity has been successfully completed in the area (as described in Chapter 2), only the correct number of each tool should remain. Although rare, there may be more than one of the same item deemed necessary. Try to avoid double-stacking like tools on the same peg hook. Give each item its own unique location. Tool boards can accommodate most, if not all, of the tool types that are required to support the flow of work in the production area.

Please note you may struggle through a couple tool board iterations before you settle on a layout that best meets the needs of the team. Once you have it set up, have those who work in the area give feedback on any recommended changes.

6. Create tool shadows: After placing the tools on the board, use paint pens from your event supply box to trace an outline of each tool on the board. Time consuming? Yes. Worthwhile? Yes. The painted outline is very visible when the tool has been removed for use or is missing. Next, make a name label for each item and place the label near the item's location on the board.

a. Name/address the tool boards: Give each tool board a designation, such as M5 (Maintenance area board number 5), J7 (Building J tool board number 7), L3 (Lamination tool board 3), etc. Label the tools with the board designation so everyone in the facility will know where a tool should be returned.

Personal Tools: Dilemma or Solution?

An ongoing challenge regarding tool management within many manufacturing companies is that maintenance workers are required to supply some or all of their own tools. Basically, the tools are their personal property and if the worker quits, his/her tools go home with him/her. While tool boards make sense for company-owned, community-use tools, what can be done with personal tools? This issue is almost always raised during each 5S implementation. One common notion is a company cannot set policy regarding the use or storage of a worker's personal tools. Even though we recognize both the fine line and emotion often accompanying both sides of this argument, as the plant manager there are some options to consider and some important questions to ask.

Although the tools are not owned by the company,

- Who owns the floor space on which the tool chest is located? The company does.
- Who owns the time spent sifting through piles of tools in personal tool chests? The company does.

Whether or not the company owns the workers tools, the business's time and money (labor hours) are directly tied to the technician's use and storage of personal tools. Disorganization and improper use of facility space causes waste, which equates to a loss of profit, which equates to a loss of jobs. When workers begin to see the value in visual boards, reduced clutter, and greater efficiency in locating needed or missing tools, they will often embrace the Lean philosophy. To reduce excess, they will begin to take their unnecessary tools home, separate their tools into categories, and designate specific drawers in their tool chests. They will label their drawers and cabinets for easy identification of the contents. Some technicians will even take this concept farther and develop "shadow drawers." Clear location and designation within a drawer allows the technician to have more organization and be able to easily see when a tool is missing.

Start applying 5S in the manufacturing facility and encourage your technicians to take the lead with their own personal equipment. For most, the idea will catch on, and those who own personal tools will begin to take pride in their ability to organize and excel.

Tool Check Cards

A great visual control idea for controlling tools is to create tool check cards that are used as an instant communicator of tool status. The 5S approach to tools is the organizational phase. It also allows for quick retrieval and quick notification if a tool is missing or in use. Once the tool is removed for whatever reason, all the shadow tells you is that, well, it's missing. But, what is the status, and who has it? Tool check cards can serve as an indicator of status. Figure 5.6 illustrates a system being used in a shipping department.

First identify the operators or workers who will be allowed to access and use tools in the respective area. If there are five people, then make three sets (as example) of five cards. You also can color code the card or make space for their names. Second, design the card so it can hang on any type of peg. Finally, laminate the cards and place them on or near the board. When someone retrieves a tool, they simply place the tool check card on the peg. Now, the entire area knows who has the tool. You can also make check cards for management or other support personnel,

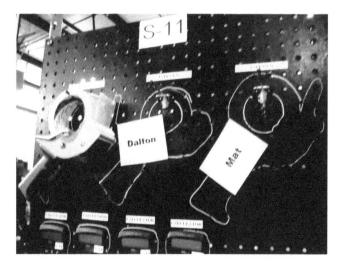

FIGURE 5.6
Illustration of the tool check card system used in a shipping department.

just in case they come into the area and need to use a tool specific to that work area. It is also smart to make tool check cards that identify if the tool is broken or on order.

Right Sizing during Set in Order

You can sum up the reason for right sizing in one word: accumulation. The more space that is available, the more space that is consumed. Whether it be a workbench, supply cart, inventory shelf, tool board or toolbox, computer desk, even an entire room or production area, if something is not the right size, more stuff will accumulate. For example, let's analyze your desk. How much space is consumed with "stuff," and how much is actual working space? Even though most of the items are necessary, are they stored in something right sized? If cubbies, bins, and other organizational compartments were to be used, would it not free up more working space? Could we then even challenge the need for the size of the desk? I think you get where we are headed here.

Right sizing first begins with an aggressive 5S implementation and requires detailed knowledge of the work content, the tools used, and the human interaction with the items in the work area. Therefore, some analysis must be performed in the current state to capture what is really going on and what is really necessary. More often than not, people tend to spread out and create a very large work area as well as large working surfaces that generate even more storage for more unnecessary items—a subconscious domino effect that can be easily controlled by right sizing.

The decision to right size stems from specific metric needs of the organization or work area. Metrics, such as floor space, travel distance, and even inventory levels, are easily measured and greatly reduced by right sizing.

Ergonomic solutions must be addressed, such as height of work surface, accessibility, or reaching high and low, but there is a close average of heights and lengths to begin with. It's easy to get consumed with exceptions and various conditions of worker heights to strive for accommodating everyone, but this quickly becomes impractical. If adjustability can be incorporated into the design or modification, this, of course, will benefit everyone interacting with the area. However, minimal tools and simple adjustments must be design requirements. Addressing ergonomic concerns is a great way to generate "buy-in" for the changes in the area and can have a large positive impact on the culture. Workers see the attention to detail and intentional effort exercised on the creation of something "right sized" and can more easily be won over with this level of detail and customization.

FIGURE 5.7
A stand-alone customized tool board.

Examples of Right Sizing

First, design your tool board or shadow board. Once all the tools and supplies are collected, begin to right size by placing all the items horizontally on a work surface with even spacing. This helps to estimate the size of the surface required to mount the items. By establishing point-of-use placement, the destination of the tool board can be determined. This could be on the side of a cabinet, built into a work table, a wall-mounted surface, a mobile cart, or a stand-alone mount. In Figure 5.7, we examine a stand-alone board.

- Notice the bins in back of the board: Right sized for specific inventory quantities.

- The items used most frequently are in the center of the board addressing ergonomics and accessibility.
- The board and frame are customized for these specific items and nothing more. No extra space is available for accumulation.

Second, design a work surface or workbench . After establishing the point-of-use items needed and assessing the surface area required to perform the task, the proper size is established and the ergonomic needs are addressed. Again, all spaces are optimized and accumulation is difficult. Figure 5.8 shows the aftermath of eliminating a large workbench and right sizing it to a smaller yet useful design.

- Spray cleaner and sandpaper have specific locations and specific quantities allotted.
- Working surface can be raised or lowered.
- Work table is mobile with locking wheels.
- Working surface only accommodates one unit at a time to reduce overproduction and support pull systems.

Finally, we explore a unique need for both floor space and accessibility. The former grinding and buffing machines were mounted to a large worktable and the consumables were kept in multiple locations. The need for accommodating larger items to be polished helps lead to this solution—a

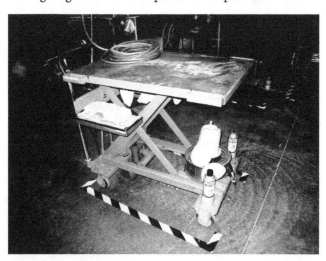

FIGURE 5.8
Photo shows the right-sized work cart that eliminates a large workbench.

FIGURE 5.9
Photo shows a customized buffing and grinding right-sized maintenance cart in a maintenance department.

suspended mount of the buffing machine. Also, the need for further mobility throughout the facility established the requirement of locking wheels and a rigid frame. Figure 5.9 shows a customized buffing and grinding cart in a maintenance department.

- Additional grinding and buffing wheels are inventoried at this location.
- Grinder tools are mounted point-of-use.
- Cart is mobile with locking wheels and is brought to the work rather than the work being brought to the machines.

Accumulation is a natural tendency and an inevitable result without right sizing. It seems like common sense. Why would we set something up that is not right sized? However, it happens much too often. Even when setting up inventory locations or bins, it is easy to take shortcuts and unknowingly allow excessive space. It can be very tempting to "settle" for a quick and fast result when brainstorming different solutions or shopping through a catalog, but remember that the best results always come from creative customization and modification. Only then will you get and maintain exactly what you need. So, if you want to truly change the mind set or even break some old thinking patterns that are costing money and efficiency, apply the concept of right sizing in your 5S initiatives.

Shine

Shine is a fancier way of saying clean. Part of your 5S implementation is cleaning, sweeping, mopping, and possibly painting everything. Your goal is showroom condition. Your visual workplace should be a showroom for customers, suppliers, new employees, and whomever will walk through. Don't underestimate the power of shine to yet again change the mind set of your people. Many times over, companies let their processes in both the office and production become quite dirty. Production processes are generally the worst, but I have witnessed some scary-looking offices.

The 5S team will more than likely be cleaning during the entire time the 5S implementation is getting done in the area. Regardless, having the team get on their hands and knees and clean puts more of a personal touch on the experience. Shine is critical, especially when it exposes safety improvements. I can remember facilitating a 5S event in Louisiana about four years ago in which certain team members were tasked with cleaning equipment. It was a heavily automated process, so there were machines all over. It was an oily and dirty environment not because of the equipment, but because there really never was any initiative to keep things clean. Nevertheless, as the team progressed, there was a situation where two operators were cleaning the dirt and grime off one piece of equipment and made the comment, "I never knew this thing was blue." Everyone found it quite funny; but as the shine portion continued, the team exposed faulty wiring, loose hoses, leaking motors, and all types of machine abnormalities. It then led to repairs and upgrades to the equipment that would never have happened if there hadn't been a cleaning.

The team was quite proud of themselves for exposing these issues. Not only did they feel good about a nice, clean work area, but they were the ones who identified major problems. Your culture gets a sense of ownership through the 5S process and team-based approach at the cleaning phase of it, which only emphasizes this theme.

Standardize

Standardize is the act of creating consistency in the manner in which the visual workplace is implemented. As you implement 5S into other areas, be consistent. I am not implying that the work areas look identical; and I also am not advocating using the same type of workbenches, tools, dollies, etc. The best way for me to relate this to you is for you to think about how visually consistent our traffic and pedestrian environments are. Imagine if a stop sign looked different in every state. If a dashed line down the middle of a road meant pass in one city and don't pass in another city.

One approach is to develop a color code system throughout the facility by identifying all the difference departments and processes to a color.

Shipping: Blue
Fabrication: Red
Maintenance: Green
Assembly 1: Black
Assembly 2: Orange
Work cell 1: Yellow

The list above is simply an example, but the consistency here is that all areas are color coded. The next part of standardization is identifying colors that have the same meaning regardless of where they are applied. This is mostly for floor items. For instance, yellow floor tape could mean items behind the workers. Green floor tape could mean garbage cans and recycle; black floor tape could mean finished goods. Whatever you think make sense.

From a psychological perspective, the actual act of standardizing an area visually has a low impact, but the overall concept has a huge impact. Being consistent with the visual workplace creates a sense of awareness and comfort. Again, going back to the concept of how standardized our traffic system is, you as a driver and pedestrian simply go about your day with little concern. It becomes very routine as you maneuver around and you don't question it.

As your production, warehouse, or repair process is visually standardized, you will create this same type of environment. More importantly, if for some reason it falls out of standard, the culture quickly knows. When the crosswalk sign is not working, people generally don't cross. They look around and begin to question inside their head why it is not working. Also as important, this quick awareness of the problem sticks out and solutions can be performed quickly. Your organization can have the same type of consistency and quick response. Try to maintain a standard approach to 5S as you move throughout your company and you will see remarkable changes in the culture.

Sustain

I often ask groups I am training, "What do you think the hardest "S" is for most companies? Nine out of ten times everyone replies, "Sustain." Sorting is a close second, but sustaining is critical. From a Lean practitioner's perspective, I look at it this way: If people are going to struggle to put the garbage can back in its location when done, the culture will struggle with everything else in Lean. It's a fact! 5S also creates accountability and a sense of structure that is needed to grow a company. A lackluster approach to organization and clutter control, as mentioned in previous chapters, clogs the brain and will not place a focus on discipline. When developing your sustaining program, here are some things to consider.

- Human will
- End-of-day cleanup procedures
- Daily walkthroughs
- 5S auditing
- 5S tracking
- Incentives

Human Will

If you implement 5S to the level I have illustrated in this chapter, you are going to need a strong sustaining mechanism in place. How you create this and the detail you place in it will dictate your success with 5S. A sustaining program must be approached with the mind set of keeping the improvements gained, identifying the next opportunity, and continually improving the program.

The first question you need to ask is will human will be enough? Do you have faith in your current leadership and other employees that they will see a collective value in 5S and they will work hard at sustaining it? Often enough, humans won't do it. Yes, there are exceptions to every rule, but most people like structure, a framework with guidelines, rules, and policies. Regardless, human will is a vital component because with structure, people have to act and commit to making the "system work." My point here is that human will may be enough to sustain your 5S initiatives; and if you feel it won't, read on.

End of the Cleanup Procedures

Basic and simple in its own definition, establishing specific end-of-day cleanup procedures in each area may be enough. We also recommend that the work area achieve a compliance level of 5S that makes it worth enacting your sustain program. Compliance is simple: Everything has a home.

Create an itemized list of the specific requirements for "resetting" the area. This is extremely important in multishift operations. It is natural for each shift to blame the other for lost information, creating defects, missing deliveries, or losing tools and supplies. Whatever it is, it is always the other shift's mistake. When 5S is correctly implemented, the above issues begin to decrease significantly. To help ensure that old problems do not creep back into the process, making requirements for end-of-day cleanup is critical. Not only does each shift reset the area for the next shift, but you begin to bridge the gap between them: 1st shift sets up 2nd, 2nd shift sets up 3rd, and 3rd shift sets up 1st.

In the eyes of your external and internal customers, they don't care about your bickering or problems; the process as a whole is responsible for the quality, output, and cost of the area. 5S can help create a more team approach to the success of the area. It will help change the way the employees in the area think about each other and their success as a group.

Daily Walkthroughs

If 5S is implemented to the aggressive visual level I recommend and the area conducts its end-of-day cleanup, then a daily walkthrough by shift supervisors will be virtually instant. It becomes clear quite quickly if something is missing or out of place. And I mean everything: information, paperwork, tools, material, parts, supplies, carts, etc. But knowing the daily walkthrough is coming, it can show the culture that this is serious and performing the end-of-day cleanup is verified.

5S Audit

Don't run away. This type of audit brings a more positive approach. I say this because the term *audit* generally has a negative connotation to most workers. This level of sustain now creates some paperwork and time to conduct properly. Not extra people, but just some responsibility to existing jobs. 5S auditing can really take your sustaining to a whole new level and each auditing approach is different.

First, with the 5Ss as a guideline, create audit criteria for each area department. Make a list of what a potential auditor would look for to ensure sort, set in order, shine, standardize, and sustain are in place. Figure 5.10 is a very good example of the types of items that are audited.

As you can see, it contains each "S" and specific criteria for each one. You also see that the scoring on the right-hand side is Yes or No; no ranges like 0–5. It is compliant or not. The 0–5 type of scoring that I have seen in some companies is too broad and the score selected can sometimes come down to a gut feeling. Make sure to select criteria that will allow for a Yes or No score.

5S Audit Form

			YES	NO
Area: Shipping				
Audit Date	# of Yes's	/18 = %		
Auditors				
Sort (Get rid of unnecessary items)				
Shipping Department is Clear of All Non-Work-Related Items				
Unused Forklifts, Pallet Jacks, and Other Lifts are Removed				
Unnecessary Garbage and Recycle Containers are Removed				
Set in Order (Organize)				
Shipping Tools are on Tool Boards and Labeled				
Forklifts and Pallets Jack Have Locations and are Clearly Marked				
Corrugated Material and Other Shipping Supplies are Organized with Kanban				
Items on Floor are Labeled and Marked				
Pick Lists are Posted and Pick Updates are Visible				
Pallet and Rack Locations are Labeled and Marked				
Scrub (Clean and solve)				
Floors, Work Surfaces, Equipment, and Storage Areas are Clean				
Garbage and Recyclables are Collected and Disposed of Properly				
Excess Pallet and Packaging Materials are Cleared out of Area				
Standardize (Tasks)				
A Warehouse Layout is Posted with Locations				
Forklift Exit Signs are Posted				
Sustain (Keep it up)				
5S Audits are Conducted Weekly				
Auditors are on a Rotation Schedule				
End-of-Day Clean Procedure is Posted				
5S Tracking Sheet is Posted and Current				

GREEN = 81% to 100% YELLOW = 66% to 80% RED = 0% to 65%
Area is 5S compliant Area meets minimal standards Area needs immediate attention

FIGURE 5.10

5S audit form for types of items in an audit.

At the bottom of the form there is a range of compliance, and the purpose for this is to say that one or two scorings of No does not warrant an out-of-compliance final score. There still is room to improve, but it keeps people engaged and shows them that small deviations are acceptable.

5S Tracking

Keeping with the model of the visual workplace, another layer of sustaining is the form of 5S tracking. It is recommended that overall scores from all areas in the company are posted visually so everyone can see how everyone is performing. This is not to point a finger, but to catch trends in areas that need more help sustaining. Often placed on a plotter-size paper or a dry erase board, it is posted in breakrooms or high traffic areas for optimal exposure. It is a great visual guideline of where to focus more improvements (Figure 5.11).

There is no hiding from the tracking sheet. It is right there, viewable to all. Without awareness, there is no change. If people can't see the opportunity, they are not inspired to change and improve.

If the people in the area simply do not want to contribute to the sustaining of their area and are just resisting, leadership needs to act and begin identifying the root cause of the problem. Many factors can contribute to the area becoming noncompliant, but if it is just attitudes, you better get on it. Letting 5S slip and allowing for disorganization to find its way back is telling your employees that, you don't care. 5S also teaches the leadership to get more committed to their business, sustaining metric improvements, and acting as examples to your investment in Lean.

Again, 5S tests the culture, but also tests those running the organization.

5S Tracking Sheet

Area	Period: 1/4/10-1/15/10				Period: 1/18-1/29/10				Period: 2/1/10-2/12/10				Period: 2/16/10-2/26/10				Quarter 1
	Day	Swing	Grave	Weekly Avg.	Day	Swing	Grave	Weekly Avg.	Day	Swing	Grave	Weekly Avg.	Day	Swing	Grave	Weekly Avg.	Quarterly Avg. to Date
Cell 1	100	100	100	100	100			100	100	100	100	100					100.00
Cell 2	100	90	100	97	100	90	100	97	100	70	90	87	80	90	90	87	91.67
Cell 3	90	80	90	87	90	90	80	87	80	80	90	83	100	100	80	93	87.50
Cell 4	100	100	100	100							80	80	90		90	90	93.33
Cell 5	100		90	95		90		90	100	80		90			90	90	91.67
Cell 6		80		80	100		100	100					90	100	100	97	95.00
Cell 7	100	100		100			100	100		80		80	100		100	100	96.67
Cell 8	90	100	100	97	80	100	90	90	80	90	100	90	90	100	100	97	93.33
Assembly	100	80	80	87	90	60	70	73	70	70	80	73		90	70	80	78.18
Packaging	80	100	100	93						70		70	70	100		100	90.00
Maint.	90	100	100	97	100	100	100	100	90	90	100	93	90	90	100	93	95.83
Shipping	100	90	100	97	80	90	100	90	90	100	100	97	100	90	100	97	95.00
Lab	Event					90		90		80		80	90			90	86.67
Warehouses	90			90	100			100	90			90	100			100	95.00

80%-100% AREA IS 5S COMPLIANT
70%-79% AREA MEETS MINIMAL STANDARDS
0%-69% AREA NEEDS IMMEDIATE ATTENTION

FIGURE 5.11
5S tracking sheet that shows how certain areas are performing.

Incentives

Not necessarily monetary, but implementing some type of an incentive program could be the ticket. Some of our clients have enacted some form of give-back for areas with the highest score. For instance, a rope manufacturer here in Washington began giving out gift cards for the work area with the highest average score at the end of every quarter. The winner also was given a 5S champion trophy and as the each quarter announced a new winner, that trophy was then moved on. Incentives may not always work, but it is something to consider and it can simply be fun and a way to bring the company closer.

PSYCHOLOGY OF 5S

5S works. It's quite powerful in improving a variety of business metrics, but it is equally or even more powerful in changing the culture. 5S can involve everyone in the company. We have facilitated 5S projects where the president is on the implementation alongside production workers. It is the tangible element of Lean that you can see after its implementation. 5S creates structure, discipline, and that first level of professionalism. It challenges the culture to change some old ways of working, and then challenges again their ability to maintain a work area. 5S sets the tone for more improvements and can have a positive impact on the mind set of your culture.

6

Making Change with Lean

Now that you have a firm understanding of 5S, you can begin to see how other Lean implementation tools can impact not only performance and company health, but also how they can change the mindset of your people for the better.

There are a diverse array of companies, products, processes, and employees. Lean, as an all-encompassing business philosophy, is applicable in any environment. However, you must understand the concepts well enough to know which application makes sense. High-volume, low-mix environments are perfect candidates for most of the Lean tools. Low-volume, high-mix environments are harder to apply certain tools. Auto collision repair shops have their own approach and administrative functions require a deep understanding of Lean office techniques. For the purposes of this book, I focus on six specific Lean implementation methodologies.

WHAT TYPE OF MANUFACTURER ARE YOU?

Lean is a very process-oriented concept. One of my first clients asked me a very interesting question when he was deciding on hiring Kaizen Assembly to assist his organization with Lean. He said, "Tell me in one sentence what you do for a living." This was more of an ice breaker because he was already familiar with Lean and he was trying to warm up to me. I answered, "I follow things." He smiled and asked back, "What do you mean?"

I explained to him that there is a process for everything. Every process is in place to complete a product or service product. All processes have a start point and an endpoint. There are hundreds, maybe thousands, of processes in an organization and they are all interconnected at some point. There is a process for assembling a product, packaging a finished good, creating a work order, developing new products, creating a marketing plan, repairing a car, hiring an employee, and the list can go on and on. My job, I told him, was to follow things and see how I can help reduce the time, steps, and mistakes that happen along the way to better serve the customer. Regardless of what your company produces, your product is made in one of four types of environments. Each one of them carries its own level of waste and has its own impact on the mindsets of your people. See where you fall.

- Process-based manufacturing
- Assembly line
- Cellular manufacturing
- Inline production

Process Based

Often considered "batch" processing, process-based manufacturing is the least ideal of the four types. I am not implying you cannot be successful as a process-based producer, but it lends itself to the highest level of waste. Process-based manufacturing facilities are broken up into work areas that basically perform different types of work. Welding, deburring, sawing, ovens, paint booths, assembly areas, packaging, etc. are separated into departments.

Process-based manufacturing generally works on one particular product or customer order to completion and creates large batches of work-in-process (WIP). The WIP is then stored in some type of outgoing staging, then transported a long distance to another processing area, and dropped to be staged yet again. Once the second process has time, the staged WIP is then retrieved and processed. And, the same cycle continues throughout the entire facility until at some point there is a pile of finished goods. Process-based manufacturing includes

- Long travel distance
- Higher WIP levels
- Longer lead times
- Long setups
- Less cross-training of workers
- Limited-to-no visibility of mistakes
- Large facility space

- Reduced sense of urgency
- Often works toward a ship date
- Limits the growth of the company
- Harder to measure productivity (Why would you measure efficient WIP production?)

Assembly Line

Assembly-based environments do not generally manufacture anything. These environments simply add parts and components together to form a finished product. A process-based manufacturer may have an assembly operation somewhere in the facility, but as a stand-alone, it manufactures nothing.

Assembly lines typically have operators standing close to each other, workstation to workstation, placing their required parts into the unit and then passing it on to the next worker. Assembly operations have been around for over 100 years, dating back to the early Ford assembly plants. Now, there are poorly designed assembly lines that create a lot of WIP, have imbalances in workload, and have longer than needed cycle times. However, there is a Lean approach to assembly lines as well that I will get into shortly. Advantages of an assembly line include

- Side-by-side operators
- No manufacturing
- Workstations are in a line
- Faster pace
- High volumes
- Better visibility of mistakes (depending on WIP levels)
- More cross-trained workers
- Easier to create a productivity measurement (product is completed at the end)

Cellular Manufacturing

Cellular manufacturing is an ideal process to convert a process-based and assembly operations. I describe cellular manufacturing in greater detail in this chapter, but here is a quick definition. Cellular manufacturing is a concept in which products that follow common processes are grouped together in a work cell environment. This is called a *common value stream*. When implemented correctly, the work cell completes products to what I call a "sellable" stage. Process-based manufacturing companies create WIP from one stage to the next, and getting to a completed sellable unit takes longer and costs more money. True Lean work cells significantly reduce the lead time required to complete products. Work cell workers are considered a team, and they work together to ensure product is flowing throughout the cell to reach a clear target. Cellular manufacturing lends itself to higher productivity and allows for the simplest form of productivity measuring.

Cellular manufacturing concepts can be applied in a multitude of industries including auto body repair, machine shops, some forms of job shops, and even in office environments.

Advantages of cellular manufacturing include

- Stations and machines close together to reduce operator and product travel
- Single-piece flow or small controlled batches
- Cells are often U- or C-shaped
- Quick changeovers
- Flexible environment for mix models
- Reduced lead times and cycle times
- Greater visibility of mistakes
- Elevated sense of urgency
- Easy output measurements
- Ideal productivity measurements

Inline Production

Inline production is kind of a hybrid of the top three advantages and its implementation requires an in-depth understanding of Lean, process constraints, and company products. Cellular manufacturing works well when you have smaller products that allow you to physically make U- or S-shaped cells where operators can move from one station to another with ease; generally within a few feet of the next operation is ideal. If your product or equipment is physically quite large and does not allow for the closeness that cellular manufacturing requires, then inline production is a very good substitute to achieve similar performance levels. Like a cell, inline production strives for the same goal: *value added to value added*. How soon can the next value-added step begin without having to create WIP, move it, and store it? Attributes to inline production include

- Stations and machines closeness to reduce WIP and travel
- Single-piece flow or small controlled batches

- Decreased time from value added to value added
- Quick changeovers
- Flexible environment for mix models
- Reduced lead times and cycle times
- Great visibility of mistakes
- Elevated sense of urgency
- Easy output measurements
- Ideal productivity measurements

Inline production and cellular manufacturing are very close to each other by definition, but do have subtle differences. Again, selecting which one depends on your understanding of Lean. To better arm you with the decision-making ability, let's dive into them and begin the Lean implementation journey. Covered in the rest of the chapter are

- Cellular and inline production
- Total productive maintenance
- Kanban and material replenishment
- Setup reduction and quick changeover
- Visual metrics and performance

CELLULAR AND INLINE PRODUCTION

Every production process has a pulse or pace. Knowing and actually calculating this pace is critical to implementing either a cellular or inline production process. The fundamental beginning step is establishing the takt time for the process. Takt is German for *rhythm* or *pulse*. In some cases, it is difficult to truly calculate this number, and I often see the concept used incorrectly. Takt time is calculated using three numbers: available time, number of shifts, and output.

$$\text{Takt time} = \frac{\text{Available time (shifts)}}{\text{Daily output}}$$

As an example:

$$\frac{420 \min (1 \text{ shift})}{35 \text{ units}} = 12 \min$$

In this example, there are 420 minutes of available time or 7 hours. If you take a typical 8-hour day and you exclude start-up, breaks, lunches, and end-of-day cleanup, as an example, there are 7 hours available for actual work. Never design a process around the full amount of time people are in the building because they are not working 8 hours. There is 1 shift and 35 units needed to be completed in a day. Takt time comes out to 12 minutes. Basically, a product needs to be completed every 12 minutes to keep up with the orders of that product family. Takt time is then used to design the process.

If you were to design an assembly line, time and motion studies would be required to get a firm understanding of the individual steps and times. For instance, if the total assembly time was 2 hours or 120 minutes, then the math would say you need 10 workstations: 120 minutes total assembly time divided by 12 minutes takt time will give you the number of resources needed. In this case, 10 workstations are needed to complete 35 units in a day. Each workstation would then be balanced in work content to no more than 12 minutes to ensure a nice flow of product. Cycle times would vary slightly between stations, depending on the work performed. However, the uniqueness of manual assembly lines is that the work steps can be broken down into greater detail.

Assembly processes should be designed to single-piece flow; so, in our example, if there will be 10 workstations, then there are only 10 pieces of WIP on the line. They do pile up in between workstations just like process-based production; all but 10 are being worked on. Any minor variations in cycle times per station will have minimal impact.

Single-piece flow environments create a sense of urgency that process-based production does not, or even batched-filled assembly lines. Operators must work as a team and they must be working together to ensure even flow. The mindset changes, and an understanding of completion and team achievement is heightened. You also will find employees becoming more proud of their accomplishments. A synergy is created and common goals of output, productivity, and quality become the mainstay of every day.

There is a larger impact to leadership in a single-piece assembly line because if quality problems arise or if a machine in the workstations goes down, the process will come to a halt. There would be no excessive WIP to pull from while reactive measures are taken to resolve the problem. Process-based production does not force a company to fix major problems. With the added inventory, what's the hurry? I am not implying that process-based companies do not try to fix major problems, but it often takes longer. Single-piece flow raises the bar on this, and you are now forced to do something for the long term. It is a pleasant evil of Lean.

To be successful with a single-piece flow process, you must incorporate standard work. Standard work is a set of operational procedures that establishes the best, most efficient, and safest work methods for all work in the plant. Each workstation must have clearly illustrated work descriptions, sequence of steps, cycle times, and visual work instructions outlining all work needed from the station. If you operate a mix model process, the standard work procedures for each product in the mix must be documented. If the assembly line begins to falter, miss output, or make mistakes, then supervisors can revert to the standard work to verify if they are following the best practice.

All required tools for each workstation must be placed on visual tool boards, material should be placed into bins or locations clearly marked with labeling and designations, and all items on the floor must have locations marked with floor tape/paint and labels. 5S is a critical piece to fine-tuning the process.

Do expect some culture shock on converting from assembly batch to single-piece flow. There is a learning curve involved, and you have to stay true to the conversion to ensure success. People will tend to revert to their old ways of working and try to batch process. Once they get into a groove, which could take weeks, they will begin to show results. You can almost compare it to surgery. There is a healing phase before final results are achieved. If you are an assembly company and as you implement more and more single-piece flow, you will see a level of healthy competition arise as everyone begins striving for excellence and the best performance. There is nothing wrong with that.

Cellular manufacturing processes are essentially assembly lines where the difference is in the physical layout change in the standard work requirements. Not every production process has assembly lines, but the company can strive for single-piece flow and WIP reduction. Many of the benefits of an assembly line are achieved in a work cell. Cellular manufacturing processes also are created to ensure product is sellable when it leaves the area. Work cells tend to be U-shaped or sometimes C-shaped so operators can move from one station to the next with ease. Calculations for cellular manufacturing are the same in regard to takt time, cycle times, and resources. However, the standard work is different.

In a traditional single-piece flow assembly line, an operator must be present in each workstation so there is no disruption. Manual assembly is considered internal work where the person must be present to perform the value-added operation. In environments where the work content to complete the product includes both manual and automated work, there is a mix of internal and external work. External work is value-added work that does not need a person to do it. A machine operator may need to do some level of internal setup work to prepare the machine for making parts, but then they can essentially walk away to perform other work while the machine runs. This is external work.

Cellular manufacturing is a better setup for this because operators can shift around the work cell, performing work as needed to keep product flowing. In a cell, you don't necessarily need a person at each workstation/machine. You still have to identify which steps each person will do, but they may need to move around to do it. This only works in a work cell.

I have seen companies change assembly line into work cells not knowing all the facts. Work cells operate better when the product and equipment are small to allow for the movement of people. This movement is wasted motion, but it is minimal if done right and the tradeoff is a massive reduction in lead time compared to process-based production. If the size of the product and machines create a work cell that takes up a lot of floor space, then it might not make sense.

INLINE PRODUCTION

Another option is inline production, which works well when product and machines are big, but you still want to reap the benefits of reduced WIP, faster throughput, reduced lead times, and a sense of teamwork.

Inline production allows for the product to go from one value-added process to the next quicker. Inline production requires the skills of manufacturing engineering and maintenance people. Good manufacturing engineers have the ability to look at processes and develop fixtures, tooling, and specialized devices that help transform it to inline production. Below is an example.

A manufacturer of mesh tubing was trying to convert existing work areas into cells or implement inline production when possible. The mesh tubing that was made came out of machines in various diameters and lengths. Diameters ranged from 1 to 5 inches and lengths up to 2,400 feet. This mesh tubing was used in a variety of applications, but commonly used in places where gripping something was critical. Pipes, handles, or tools could all use this product, and the company produced millions of feet of it a year.

The mesh tubing would come out of the machines and slowly go into a large tub on wheels. It took hours for the product to be made and, within a 2,400-foot run, there could be 24 orders of 100-foot lengths. Customers bought it in 100-foot bags. Once the tube was full, it was then rolled into another area (process based), where it sat in queue to be bagged. The bagging stations would place the mesh tubing overhead, and then another machine would insert it into a bag. With a length measurement device on the machine, the operator would watch until it hit 100 feet, stop the machine, cut the tubing, and move onto the next bag. Once a pile of bags was filled, the operator would place a header card on the bags and place them into a box.

My engineer and I watched the mesh tubing process one day and presented an idea. Why can't they replace the tub with the plastic bag that is used in the bagging operation? Why can't product flow directly into the bag rather than a tub? This would be *inline production*. They liked the idea and ran with it. They placed a length measurer on the machine to ensure they knew when they reached the desired length to cut. Because we also were implementing cellular manufacturing, part of a person's standard work was to periodically go by and check the measuring devices to see if a cut was soon needed. Maintenance and manufacturing engineers designed and fabricated a turntable that could hold up to five plastic bags in a fixture to keep them open. As a bag was complete, it was removed and the next bag went into place. While the machine continued to flow mesh tubing into the second bag, the operator would place the header card on the bag and finalize the packing (external). This truly simple improvement not only allowed the cell to begin completing products to a sellable level, but also reduced the overall lead time by two days. Travel time and wait time were eliminated.

Really, the best take-away from this section is that if you challenge the way you transport and handle WIP and slowly reduce the distance and the quantity of WIP, you will see dramatic reductions in lead time. You also will be able to change over more quickly, have better visibility of problems, and position yourself better for growth. Single-piece flow assembly, cellular manufacturing, and inline production is a step worth taking, not just for the process and the company, but helping mold your people into achievers and forward thinkers.

TOTAL PRODUCTIVE MAINTENANCE (TPM)

TPM is a company-wide approach to improving the effectiveness and longevity of equipment and machines. Depending on how automated your production processes are, TPM could play a major role in your Lean journey. TPM is also a critical component of the success of any cellular manufacturing operation because, to maintain flow and uptime, equipment must be working at optimal levels with little downtime.

Regardless of proactive measures, random breakdowns may occur. The goal here is to minimize their occurrence where most of your maintenance is conducted during planned downtime. Part of a comprehensive TPM program is creating accountability and ownership as well at all levels to help change the mindset of the machine operators and maintenance. Both are responsible for maintaining equipment, and each plays its own role. We often see a laissez-faire approach from maintenance departments in traditional manufacturing companies where machine maintenance is mostly reactive. Without an organized and structured TPM program, maintenance staff can develop a feeling of invisibility in which they are the lifeline of the company. Without them, the company will fall apart. Wrong. They play a vital role in the overall health of the company, but they must work according to certain standards and guidelines just like we place on production workers. Benefits of TPM include

- Reduced downtime
- Reduced breakdown costs
- Reduced spare parts carts
- Reduced defect rate
- Reduced lead times
- Improved on-time deliveries

For the purposes of this portion of the chapter, I am going to break down TPM into the following categories:

- 5S maintenance
- Equipment baselines and one-turn method
- Operator preventative maintenance (PM) requirements
- Maintenance department's PM requirements
- Visual TPM boards

5S Maintenance

Another reason why maintenance workers may walk the path of reluctance is that many companies do not properly invest in the department because they do not create product. The department is an expense and a nonvalue-added function. And, on top of that, the department is required to provide adequate support to the facility. As a starting point, 5S must be implemented into the maintenance department at the level of visual detail anywhere else. You can look at it this way: How can we expect maintenance workers to support an entire facility if the environment they work in is cluttered, unorganized, and not visual. How much time do they spend sifting through tools, looking for spare parts, searching for work orders, etc.?

Maintenance departments are notorious for holding on to things for years, and the area can become a dumping place for unneeded items from other areas. We often see excessive purchasing of supplies and parts, including uncontrolled vendor-managed inventory programs. 5S and visual control can be very beneficial in reducing inventory, creating visual signals for ordering, and improving the productivity of the department. I discuss visual material handling processes when I describe Kanban in the section below (Kanban and Material Replenishment). Follow the guidelines from the previous chapter and create a visual maintenance department.

Baseline Equipment and One-Turn Method

As you progress through maintenance with 5S, start thinking of where you will begin creating structured PM requirements for operators and maintenance. Often, one of the best places to start is the actual maintenance department. Brake presses, welders, saws, drill presses, and CNC machines, for example, are commonly found in maintenance. Sometimes picking an area with common equipment, such as a saw room, is a good place to start. Focus on those machines and refine the TPM program with them, and then roll it out.

Equipment baselining is often called an *abnormality assessment*. Use criteria of some kind to evaluate the current state of the machines. This assessment is to identify where certain components of the equipment would need repairs or modifications to bring them up to a standard for maintaining. Common elements of an abnormality assessment would evaluate the following:

- Minor flaws, parts, and vital components
- Basic conditions not fulfilled
- Inaccessible places

- Sources of contamination
- Sources of quality defects
- Unnecessary items
- Unsafe places or conditions

Machine baselining is an example of a form (see Figure 6.1) we use at Kaizen Assembly and provide for our clients to conduct the analysis. You may have a similar form and criterion or even have one that works well now. Use this form to conduct the assessment, and then schedule the repairs or modification.

In conjunction with the abnormality assessment, I also recommend that you conduct a one-turn method analysis at the same time. The one-turn method methodology is a concept of reducing the quantity and time

Kaizen Assembly

			Yes	No	Priority 1,2,3	Schedule Date	Percent Complete	Comments
Minor Flaws 1	A	Missing Paint						
	B	Missing Decals						
	C	Scratches						
	D							
			Yes	No	Priority 1,2,3	Schedule Date	Percent Complete	Comments
Parts and Vital Components 2	A	Cracking						
	B	Damaged Threads						
	C	Needs Repair or Replacement						
	D							
			Yes	No	Priority 1,2,3	Schedule Date	Percent Complete	Comments
Basic Conditions not Fulfilled 3	A	Lubrication						
	B	Leveling/Upright						
	C	Loose Electrical Connections						
	D	Alignment/Adjustment						
			Yes	No	Priority 1,2,3	Schedule Date	Percent Complete	Comments
Inaccessible Places 4	A	Special tools reqired						
	B	Poor Visibility						
	C	Awkward positioning						
	D	Heavy Lifting						
			Yes	No	Priority 1,2,3	Schedule Date	Percent Complete	Comments
Sources of Contamination 5	A	Seals/Gaskets						
	B	Fluid replacement						
	C	Cleanliness						
	D	Filters						
			Yes	No	Priority 1,2,3	Schedule Date	Percent Complete	Comments
Sources of Quality Defect 6	A	Needs Calibration						
	B	Machine Upgrades						
	C	Improper Materials						
	D	Insufficient Tooling						
			Yes	No	Priority 1,2,3	Schedule Date	Percent Complete	Comments
Unnecessary Items 7	A	Tools						
	B	Personal Effects						
	C	Obsolete Information						
	D	Spare Parts						
			Yes	No	Priority 1,2,3	Schedule Date	Percent Complete	Comments
Unsafe Places or Conditions 8	A	Ineffective Panels/Shrouds						
	B	Insufficient Signage						
	C	Maintenance Safeguards						
	D	Pinch-Points						

FIGURE 6.1

One Turn Method Assessment				
Machine/Equipment	Department/Area	Location on Machine	Purpose of Application	Recommended Change

FIGURE 6.2
Form evaluation.

associated with turning operations. More specifically, equipment often needs to be dismantled in preparation for more time-consuming preventative maintenance activity. It could be the removal of doors, panels, enclosures, guards, etc. Often these components are secured with studs, nuts, bolts, and screws that all require a tool of some kind. A one-turn method analysis would identify where these turning functions could be transformed into quick disconnect and reconnect devices. By doing so, you significantly reduce the time required to remove the components and then reinstall them after the PM is complete. Figure 6.2 is a simple form we use when conducting these evaluations. I will cover one-turn methods again when I describe quick changeover, but I am sure you can think of machines and equipment in your plant that would benefit from this conversion.

Operator PM Requirements

When establishing either operator or maintenance-level PM requirements, it is best to use a three-category approach. Most PM will fall under the following approaches:

- Proactive PM
- Preventative PM
- Predictive PM

Proactive TPM

Being proactive may mean a list of simple-to-moderate level checks, such as

- Checking settings
- Cleaning and inspection
- Lubrication
- Abnormality detection
- Precision checks
- Triage maintenance
- Preventative TPM

Some items are designed for periodic replacement. Based on suggested original equipment manufacturer (OEM) intervals, we can set replacement times in our TPM plan. Some examples may include

- Replacing parts (e.g., worn out belts, filters)
- Replacing and refilling vital fluids (e.g., lubricants, cutting fluids)
- Recalibrating devices
- Developing new procedures for PM

Predictive TPM

Having a predictive system will rely upon history or an established time frame for:

- Filter or dust collectors
- Timing belts
- Hours on machine
- Air compressor fluid

Predictive TPM suggests the ability to predict, based on both collected data and OEM recommendations of a more accurate timing of part replacement or refurbishment, and not just those items designed for periodic replacement. Oftentimes, load-bearing surfaces wear and need repair. A predictive approach would restore the worn surfaces prior to producing nonconforming products.

Although not all companies involve operators in providing preventative maintenance to the equipment they operate; those who do invite operator participation are finding tremendous benefits from doing so. Although not difficult to perform, these maintenance activities invite operators to become stewards of the equipment, which brings to light machine issues sooner, and it also relieves the maintenance team from some of the daily workload. This is work that does not require oversight from the maintenance team, which means that the machine operator can complete these actions autonomously. This is true because they have been trained and certified to complete the work. Typically, these are activities that occur more frequently (daily or more often), they follow a written procedure, and the steps are usually relatively simple. One consideration is that the operator must be allowed sufficient time to perform the PM effort. Normally, the ability to spot early signs of trouble more than makes up for the time required to complete the activity. Below is a growing list of operator-level preventative maintenance guidelines and PM content currently being performed by companies that embrace the TPM concept.

Work performed more often (daily):

- Follows a written procedure
- Follows a written frequency
- Usually simple
- Must be allowed time to perform
- Cleaning equipment
- Checking fluid levels
- Checking equipment settings
- Filling vital fluids
- Identifying problems for maintenance

Maintenance Technician

With operators shouldering some of the PM work, it allows the maintenance team to remain focused on the preventative, predictive, and proactive activities. Although most machine manufacturers provide a list of recommended PM activities, most discriminating users find that additional items are necessary to ensure the machinery continues to perform as desired. These standard PM steps, which are based upon both a written schedule and a written procedure, are performed less frequently, such as weekly, monthly, or annually, to include

- Machine repairs
- Machine or equipment teardown

- Machine modifications
- Replacing parts
- Work completed during planned downtime
- Follows a written procedure
- Follows a written frequency

Depending on your current situation, you may have these elements of TPM already in place. *My* recommendation is to evaluate the current state and see if changes to either operators or maintenance should be done. There may be PM that maintenance is performing that truly should be done by the operator, and vice versa. Make sure that you identify the required work, sequence, parts, tools, supplies, and time required to do the work. Then create a procedure that will standardize the requirements.

Visual TPM Boards

Developing PM requirements for both operators and maintenance will help create better teamwork between the two departments as they both will have a stake in each performing their work. A critical piece of the TPM program is the system you create to ensure the work is being completed and that there is a nice flow of communication back and forth. I recommend you create a visual TPM board to act as the accountability structure for this purpose. Figure 6.1 is an example of a visual TPM board.

A visual TPM board goes way beyond a sign-off system. The board contains vital information about the overall PM requirements of the machine or area. The board should contain the following:

Calendar: A daily, weekly, monthly, or quarterly calendar that shows the frequency when either the operator or maintenance technician is required to perform the work.

PM Procedure: As seen in Figure 6.3, the actual procedures are posted on the board for reference, and the specific procedure in the manual is listed on the calendar. This is how a person can make the connection between what to do and when.

Magnets: These work well to visually display that the work has been performed. As the operator or maintenance technician completes the work, he/she simply slides the magnet to represent its completion.

FIGURE 6.3
An example of a visual TPM board.

Make sure to assign different colors of magnets for each department and shift so there is more accountability.

Machine Status: It also is good to post a visual indicator of possibly green, yellow, and red to communicate machine status.

Departmental Communication: A place to write comments so each department can communicate as to what it will do. However, I like to see some type of color card that can be filled out and placed on the board with information about the problem.

A company-wide TPM program can have a profound impact on both production performance and workers. Bridging the potential gap between production and maintenance and creating a culture of teamwork and alignment will help change the mindset for the better. It will reduce the amount of finger-pointing and, as a leader, it will show your commitment to the long-term health of your assets.

KANBAN AND MATERIAL REPLENISHMENT

Kanban is the Japanese word for *signal*. Visual signals are a critical element of any Lean journey as visual signals reduce confusion, time, motion, and transportation. To help resolve the psychological dilemma that comes with excessive inventory, develop a Kanban system where applicable. There are a large variety of Kanban systems, and each one must be tailored to fit the process. From a production process perspective, you want to keep your value-added operators focused on building product and running machines. However, they do require material, parts, subassemblies, WIP, and supplies to keep this going.

Material handlers are often deployed to act as a support function to deliver the necessary material and supplies. Unfortunately, the process of material handling is never created to work efficiently. Material handlers are looked at as indirect labor, so little focus is spent on their process. Streamlining this nonvalue action is critical to the success of the production line and especially in cellular manufacturing or inline production situations. Because these types of processes are producing a high mix of product, keeping inventory low but available is key. You can think of a material handling process like a pit crew in a professional auto race. The pit crews are the material handlers that service the racer and car as it comes in for vital assistance. Pit crews work within specific standards and procedures, and follow the same action on all pit stops. They are highly efficient and must work in unison to ensure the racer is serviced and back on the track quickly. As a team, they win the race together. A production line is no different. Kanban cards are an effective material handling system that can be applied in finished goods, production, and in the stockroom.

To begin, identify all the parts, material, and supplies in the work area/ stations. Then identify the maximum amount you want to have in the area. This amount is generally based on time, such as a one-day supply. Identify the minimum amount that will act as a safety stock, maybe a half-day supply. The difference will be the reorder quantity. Make sure to retrieve the part description and part number as well. You can take the same approach with supplies or consumable-type items.

The challenge is determining the correct amount to keep on hand. For example, how many rags should you stock? How long will two boxes of 100 rags each last? How often do your shop personnel go through rags? With a little math, you can estimate initial usage and then monitor actual usage.

The stockroom contains vital supplies and necessary material to support the production operation. Some supplies may be used to support production and some only for general use or facility maintenance. Regardless, all supplies must be set up on a Kanban system to monitor usage, keep costs down, and reduce carrying needs.

Here is a simple approach:

1. Identify all the material and supplies needed in the stockroom.
2. Separate them into categories (i.e., fluids, adhesives, sanding, general maintenance, electrical, hardware, etc.).
3. Identify the amount needed on hand at all times. Ask this question: How much is one-week's worth?
4. Identify the reorder quantity: one bottle, two boxes, etc.
5. Decide where the supplies will be located in the parts room.
6. Implement 5S for the supplies so that each item has a home location, regardless of size.
7. Label the items.
8. Print out Kanban cards and place them near the item (Figure 6.4).

Notice that the Kanban card includes the item description and part number, the minimum on hand, the reorder quantity, and the Kanban card number (in the upper right-hand corner). When the minimum quantity is reached, that is the signal for replenishment. Sometimes, another visual signal, such as a red dot under the product, can be used to indicate that the card should be submitted for replenishment. When the "signal" is present, the card should be placed into a Kanban card bin. The card is now the signal to order the supply. You may want to wait until a collection of cards is accumulated and then place the order. The Kanban card number

---

Circuit Bracket K # 10

PN #: 56-8873

Max: 35 #Min:#20

Reorder: 15

Location: A1

FIGURE 6.4
Illustration of a Kanban card.

is to reference the card in possibly a master list of Kanban cards to help keep track of them.

A Kanban system will ensure that production workers do not run out of material and have a nice fluid supply at all times. Because one of the purposes of Kanban is to challenge inventory levels in production, you will find it leads you to challenge stockroom levels. This system helps you see the gap between what you buy and what is used in production, and the same questions now can be asked about suppliers. You won't have the same control as you do in setting production-level inventory, but you still find the same opportunities in the stockroom.

SETUP REDUCTION AND QUICK CHANGEOVER

Reducing setup times will become an ongoing battle and developing quicker changeover concepts will ensure that you can continue to offer a variety of products. Large lot processing is a dangerous business, and many companies think that reducing the number of setups is important. Unfortunately, this is not true Lean approach. Yes, setups are nonvalue added and, at certain steps in the changeover, there is downtime. If you offer a mix of products, then frequent changeover is the ideal state. So, it is not about reducing the quantity of setups, but more about reducing the time required to do them. Here are some problems with long setups/large batch processing and what is created:

- Excessive inventory
- Delays for customers
- Delays for consuming processes
- Adds storage and handling costs
- Requires floor space to store
- Hides quality errors in large piles
- Reduces flexibility in production

If you are opting to implement cellular manufacturing, quick change-overs are crucial to keeping product flowing at the desired rate and quantity. Even if cellular manufacturing is not in the cards at this time, you should be constantly finding ways to reduce setup time. Here are some tips:

- Layout
- Tool placement
- Visual setup boards
- Material handling
- Intermediate jigs and one-turn methods

Layout

Conduct an evaluation of where the machine operator is walking to retrieve vital resources, such as tools, parts, documentation, tooling, and fixturing. Identify what items are in common use among all changeovers; then implement 5S and the visual workplace to make clear locations and designations within reach. You can construct a "spaghetti" diagram that essentially uses a layout of the work area, and you can trace where the worker is going to retrieve items. It will identify how you can change the layout of the area and even the individual workstation for reduced motion and transportation. However, the key here is to not only move it closer, but also to incorporate 5S as the organizational system.

Tool Placement

Certain tools may be needed to remove paneling, secure tooling, or perform some minor adjustments on the machine during the changeover. First of all, try to standardize your tools so that you reduce the quantity on hand. I witnessed many times where an operator was securing a fixture into a machine and was required to use multiple types of tools. The hardware is not standard and this forces nonstandard tools. The initial improvement is to create a 5S shadow board where all these tools are placed at point of use. Sure, it's a good first step, but I would challenge the company to standardize its setup tools. One more step: identify the exact location of tools used and place them right where the work is performed. Sometimes this approach is even better than an all-encompassing shadow board. You still have to implement 5S in each place, but now the tool is truly at the point of use.

Visual Setup Boards

Lack of visibility can really create problems for a production process and a company as a whole. I remember that sometime ago I received

DATE:_____ Packaging Start-Up Board							Work Order Sequence #1
Department	Work Order	Item	Responsible	Completion Time	Complete	Issues	
Maintenance		Turn On	Bob	6:00 a.m.			
		Setup Codes	John	6:30 a.m.			
							#2
Production		Work Order Ready	Jill	7:00 a.m.			
		Material Ready	Ryan	7:15 a.m.			
							#3
Quality		Spec Sheets	Sam	8:00 a.m.			
		Metal Detector	Sam	8:30 a.m.			
		Audit	Jenny	8:45 a.m.			

FIGURE 6.5

Visual setup board designed for a tortilla-making company.

a phone call from a client in Dallas, TX, who owned a tortilla-making company. We had been working with them for some time on 5S and TPM. The owner had recently been certified through our Lean Champion Certification program and was now ready to begin reducing setup times.

He explained over the phone that during the training I had mentioned the importance of developing visual setup boards that provided vital information about prescheduled setups. He asked me for a design concept for his packaging line.

Visual schedule boards should be posted in any work area that has machines that require changeovers on a regular basis. It is important to provide visibility to machine operators on what required setups are coming. Figure 6.5 shows the visual setup board we created for him to try out and modify as needed. Any visual setup board should contain the following items:

- Order of work
- Required material/parts/supplies
- Required documentation
- Completion time for items
- Who is responsible
- Status
- Completion check-off

Most of the work that is outlined on the board should be performed while the machine is nearing completion on another run of parts. If the operator can truly walk way and perform this work externally, then that is the best-case scenario. If not, it's best to incorporate your own setup pit crew to use the board as a guideline to gather the required items for the machine operator. This board acts as a guide throughout the day to ensure there is no unplanned downtime due to missing parts, material, people, documentation, or tooling.

Material Handling

We worked with a machine shop here locally in Bellingham, WA, that incorporated a setup pit crew. There was an area that was designed to hold all required parts, material, documentation, and tooling in one place. Each set of necessities was organized by machine type, and the company also implemented 5S in the area. Everything had a place and there were visual designations on all items addressing each item to the intended machine.

A visual schedule board was posted in the pit crew area and the crew used the board on a regular basis to prepare upcoming machine runs. Signals were placed in the machine centers where operators communicated with the pit crew on work completion and when to bring in the next run's required items. Specialized carts were ma'de to allow for quicker delivery in and out of the pit crew area. Kanban cards were placed in the pit crew area as well to help trigger the need for common material being consumed in both the machine centers and the pit crew area. Although no process is perfect, it still worked fantastic.

Intermediate Jigs

A simple concept and applicable in many situations, an intermediate jig is a device that allows for a quick changeover to take place when multiple pieces of material are needed. A good example would be in a robot welding department at a wood stove manufacturer. Multiple fixtures were used, depending on the type of firebox being welded. A fixture would be placed into the robot, and between four and eight pieces of steel were placed individually on the fixture. Each piece required a certain orientation and connection to ensure proper placement. Calibration tools were used to ensure accurate placement. The doors would close and the operator would select the program to run the robot—long changeover.

After evaluating the changeovers for a few days, the company started to challenge the way it performed changeovers and began thinking of intermediate jigs. Two special universal platforms were created that allowed the robot operator to place the pieces of steel into one universal platform while the other one was in the robot welding. The upcoming firebox had all the required pieces in place and was calibrated for an accurate fit while the machine was running. Once the machine was off, the universal platform in the robot slid out and the next one slid in. Changeover time was drastically reduced and it allowed them to change over quicker to other firebox models and produce more product in a given shift.

Intermediate jig creation generally takes the expertise of the operator, a manufacturing engineer, and maintenance technician to develop, but the rewards are worth it.

VISUAL METRICS AND PERFORMANCE

The final section of this chapter is dedicated to describing the importance of visual communication with regard to performance, goals, and objectives that are needed to support a visual factory. The following items are discussed in detail:

- Facility performance
- Metric communication boards (production level)
- Production control boards

Facility Performance

One of the first steps in creating visual communication is to relay important company information as a whole to the entire facility. Often, manufacturers have multiple facilities throughout the country or even the world. They should try to keep the information specific to the facility so there is more of a connection to how the employees of the plant are affecting its performance. This does not imply that overall corporate performance is irrelevant. This can be contained in some other form of communication.

First, your plant needs to decide what information makes sense to communicate. Some plants keep performance information out of sight of the employees due to the communication of bad information, such as reduced orders, poor deliveries, poor sales, etc. To be honest, this is exactly what should be communicated. Obviously, we want the information to be positive, but showing performance trends is healthy. Below are the company metrics that should be communicated visually on a company communication board in places like breakrooms.

- Sales
- On-time delivery (OTD)
- Productivity
- Quality
- Safety

Sales

Communication of ongoing sales information is very smart. Employees can see how the sales of the products they touch and build every day are moving into the hands of consumers. As sales increase, the plant will gain encouragement that the revenue stream is growing. On the other hand, a decrease in sales is equally important to communicate to the employees. The purpose of this category is to show sales trends to everyone so they can have a sense of the money coming in or going out. Your challenge is to decide how much sales information should be posted visibly in community areas, such as breakrooms. Regardless of your approach, sales information is at the top of the list.

On-Time Delivery (OTD)

Plant performance on OTD must be communicated on the same board. This metric can be a little tricky because some plants tout 100% on-time delivery, but the reason for having such a high level is because of excessive finished goods due to the act of overproduction. Of course, this does not imply you shoot for a 70% OTD. First, you have to decide how you will measure OTD. Do you wash your hands clean after it leaves the facility? Are you responsible for the shipping method and timing? Do you constantly change the delivery date because of engineering changes? Is it over when the product shows up on their receiving dock? These are the questions you need to ask and decide how you want it measured. There should be one delivery date, and that is the mark you shoot for. Percent on-time delivery is the metric here, and an industry performance of 98% is a good mark.

Productivity

The next metric that should be communicated is plantwide productivity. This is taking into account all the individual work area, work cells, or assembly lines productivity and developing a cumulative score to show overall plant efficiency. Like any metric, the selection of the measurable is important. Productivity can be measured in a variety of ways. A productivity measurement requires some kind of input: labor dollars per unit, the distance a product travels per person, pound per machine, bag per person, etc.

Productivity is improved when products are manufactured with less effort, fewer workers or hours, less equipment, and less use of utilities (overhead). If you are using labor dollars per unit, you are essentially measuring your ability to overproduce. Are those units you are making and measuring at labor dollars per unit heading straight to the finished goods shelf where they will sit until an order comes in? A better productivity measurement could be sales per head count or labor dollars per unit sold. Regardless, like all the items on your facility performance board, you should show productivity and its current trends.

Quality

Quality is another key measurable for a factory. Surprisingly, we have come across companies that do not measure quality—a big mistake. Quality should be measured internally and externally because there truly is a difference. Internal measurements of quality could include parts per million (ppm), rework hours, rejects, scrap, or nonconformances. External measurements of quality may be customer complaints, warranty claims, cost of resolution, and cost of service calls. I would recommend you communicate both internal and external quality information to everyone on this communication board.

Safety

To start, favorable safety results are driven more by a culture with an "attitude of safety" as compared to expected compliance with the company's safety policy and regulatory mandates. Either way, a manufacturing company's safety goal is normally to provide a safe workplace where all employees and visitors are free from the danger of incidents or injuries. Whereas an incident may be considered a near-miss, an accident is an event when someone is injured or killed. The impact on both the employer and employee can be catastrophic. Company leaders who find ways to create a culture supportive of an attitude of safety will benefit everyone.

Whether or not a report is required to be posted, sharing information regarding accidents and incidents is a healthy contributor to the visual factory.

Following are some items to report:

- Accident-free days
- Costs of lost workdays
- Number of accidents
- Number of incidents
- Hazardous material storage compliance reports
- Fire inspection reports
- Corrective action taken to remove the cause

Although posting these reports will help raise the awareness of the employees who read them, ultimately ongoing training and positive reinforcement will increase the level of importance in one's mind to work safer.

Metric Communication Boards (Production)

It is important to take the concept of the facility performance communication board down to the production and departmental level. Communication boards should be constructed and placed in each work area, communicating visually the performance of the individual work area. Although plantwide OTD is important to know and improve, it does not necessarily help the individual work area improve its own OTD.

Metrics at the area or cell level need to be relevant to that area. What is the area productivity and quality? What is the goal for output and delivery to the *next* process? For example, if a product is partially manufactured in cell 1, and then has to travel and be processed in three more areas, the plantwide OTD really serves no purpose. We like to call it the work-in-process date, or WIP date. When is it due to the next consuming process? That is their immediate internal customer. When developing area or departmental metric boards, try to keep them as standard as possible so each one is conveying the same metric. We recommend the following information:

- WIP OTD
- Productivity
- Quality
- Safety
- Attendance
- 5S scores

Production Control Boards

Imagine yourself as a coach on the football field. Coaching requires constant decision making and the creation of plays during each quarter, which will drive the team to score and, ultimately, to win the game. Every game is coached on a per-play basis, and you want to avoid going into overtime at the end of the game, as that will fatigue the players and may cause decisions that are not part of your playbook. Imagine how difficult it would be to coach a game without a scoreboard to help guide your decisions and your team toward success. Every decision is based on the current situation or status, and with the overall objective of winning. Your facility is no different. Each day, you and your people must make decisions to ensure that the work is done in time for customer pick-up.

Production control boards can be one of the most valuable management tools for increasing output. Of course, in order to be effective, the information on the board must be accurate, and, when it is, the results are absolutely AWESOME! A production control board monitors production progress with real-time information, allowing everyone to see if the process is producing the required amount of work. To illustrate this, let's think about an assembly operation where the line is required to produce 20 widgets per hour and associated takt time has been set to accommodate that.

If the required output is not being met, you can see this on an hourly basis, not just at the end of the day when it is too late to do anything about it. Assembly plants also have used takt time to monitor progress by counting how often a widget would have to come off the line; whatever makes sense for each company and process. Each production control board must be designed for the needs of each process; not a single template exists that would be applicable to all and any processes.

There is a long list of Lean methodologies, and the ones I described in this chapter should not only give you a great start at improving the performance of the company, but also change the way people think about their work environment. Don't forget that 5S is a Lean methodology; I just felt compelled to dedicate a chapter to it. How you mix and match, blend together, and customize your Lean implementation will be different than another company. Don't be in a hurry with Lean; it takes to time learn and implement, and you may make mistakes along the way. It's all part of the learning process. Don't settle for the status quo—change and progress ahead of the competition.

7

Keeping the Lean Fire Going

Lean will constantly be testing your abilities as a leader from the very beginning. You will find balancing the day-to-day operations of the company and incorporating Lean to be challenging at times. As the years pass, depending on your drive, it will become easier and a way of life. You want to make sure during hiring phases of the company to bring people on who have some level of Lean experience. Previous Lean training would be a minimum requirement.

It is equally important to consider looking at how to encourage and continually invest in current employees. Developing some type of incentive and solid internal training and progression programs is vital, not only for existing employees, but the newcomers too.

Organizations that truly embrace Lean and continue to fight through the battles of culture change find ways to return the favor to the people in the company who made it happen. You are aware now that Lean is a company-wide approach to continuous improvements and, as time goes on, more and more employees are involved in the process. As the company begins to see the financial rate of return on its Lean investment, there should be a way to trickle the additional profit back to the employees.

GOALS

Implementation of Lean into an organization can have a positive effect on cost. The rate of return will always differ from one company to the next as every approach to Lean is different. Some companies have seen savings of

$50,000 to $1 million in the first year, and as high as $500,000 to $4 million after five years of doing Lean. The smart ones put some of that savings back into the company and in the employees' pockets. Really, a better way to describe the financial return is opportunity cost. To actually save money, an element of existing cost structures would need to be reduced. If material, manpower, and facilities cost are not truly "saved," then it becomes opportunity cost or better utilization of the cash flow. Often this is difficult to bite off for leaders looking to save money; but unless you start laying people off or start cutting inventory levels (bad habits), your financial return is in smoother cash flow.

I have always been a firm believer in giving back when it comes to Lean. However, the company should develop an incentive system that is based on goals. As company decision makers, you want to create an exciting energy around Lean. Part of this excitement is placing goals in front of the company that everyone will strive to improve. When improvements to these goals are obtained at the end of the year, employees are rewarded financially. These goals are the key shop-floor metrics outlined in the company's strategic purpose as described in Chapter 1. The metrics include

- Productivity
- Quality
- Inventory
- Travel distance
- Floor space reduction
- Workstation reduction

Annual targets must be set by management and then employees in the company can go after them.

I discussed in previous chapters the importance of finding a balance among cost, quality, and delivery. Some companies use these three indicators as the company metrics. As an operator, they really mean very little. Not implying operators do not understand the concepts, but these three are far out of their range of responsibility. Even engineers may not know what needs to be done to improve cost, quality, and delivery. My point here is that productivity, quality, inventory, travel distance, floor space reduction, and workstation reduction have much more definable parameters. These six shop-floor metrics are directly connected to cost, quality, and delivery. So, company incentives or bonuses are contingent upon meeting or exceeding the goals for each metric.

Kaizen teams will be created and multiple kaizen events will be scheduled to implement Lean principles into the company's various departments. Employees are now more encouraged to participate in kaizen and try to better themselves and their own way of working. Continuous improvement efforts will accelerate, knowing there is an incentive and goal at the end. Use these goals as the catalyst to vastly accelerate the Lean journey. Although Lean will become your new way of conducting business, and employees must get engaged, I would like to see a payback. If an organization books a cost savings of $500,000 in the first year, then return some of that to the people who made it happen. But, it starts with targets and goals, and make sure to raise the bar a little higher every year.

PAY-FOR-SKILL PROGRAM

As you can tell from Chapter 6, I am an advocate of employee training. When meeting with clients during the Lean strategic planning sessions, I schedule time to talk with the line operators and get their opinions on the journey ahead. The most common topic that is raised during the line operator session is training. More specifically, formalized training for new employees. To avoid redundancy, I am not going to outline new employee training too much further, but any good employee training program must be backed by what is called a *pay-for-skill program.*

Pay-for-skill programs provide financial incentives for learning more jobs in the company. It is, in essence, a career advancement program for production workers to encourage cross-training on the production floor. For the company, it helps encourage operators to become more flexible and skilled so the organization can adjust to seasonality and changes in volume. For the worker, it outlines a clear path for advancement and growth in the organization.

Each process, assembly line, or work cell should have its own pay-for-skill structure. Although some processes may be very similar, there may be subtle differences in the jobs and work in each process that warrants a custom-made program. The cross-training matrix for each process is one of the guides in establishing this type of program, but there is a lot of detail involved.

The number of progression levels in a pay-for-skill program will vary and truly depends on the process. As you develop your levels, I recommend the following criteria:

- Number of jobs
- Experience
- Attendance
- Kaizen and kaizen event participation
- Quality errors

Number of Jobs

Each level should contain a certain number of jobs for which each operator must be certified. For instance, to fulfill the "number of jobs" criterion, an operator must be certified in three workstations. Becoming certified in just one workstation does not satisfy the criterion. If an organization follows the rules for new employee training and cross-training, then the company knows the operator is experienced. Three workstations is a suggestion, but I have seen some companies require certification in five or seven. It's up to the company and the complexity of the manufacturing process.

Experience

The experience criterion should be similar to the timeline for becoming a certified operator after completing the novice portion of the cross-training program. The fact that there is a time frame during this phase of operator development, including the number of jobs required in the level, there should be no question about the experience level of the worker. The guidelines for experience are already in the cross-training program.

Attendance

Absenteeism, tardiness, and turnover problems can create major problems for a company. It is important to develop a criterion for attendance in the pay-for-skill program to not only encourage people to come to work, but to also provide an incentive for doing so. This is sometimes difficult for management to swallow because good attendance is simply an aspect of having a job. I worked for a company that had 55% turnover among production workers, and that did not include the number of people who were continually missing work or showed up late.

Most organizations have defined guidelines for attendance generally based on a point system. For example, an operator may be allowed 10 points every year and, as they miss work, arrive late, or call in sick, points are removed from the original 10. A missed day could be one point, showing up late could warrant a quarter-point deduction, and maybe calling in sick is worth a half point. Some of you reading this probably have similar structures.

If an employee has exhausted all of his/her points before the end of the year, he/she is given a verbal warning. The operator is essentially on "probation" and cannot miss any more days; he/she receives a written warning. Of course, any infractions after this point would result in termination. The attendance point system will differ from one company to the next.

When developing the attendance criterion in a pay-for-skill program, there should be tighter parameters. To move on to the next pay level, attendance should be nearly perfect, with some exceptions. Simply staying with the provided points does not warrant promotion. One approach is to allow progression to the level if any missed days or tardiness with notice is acceptable. People miss work for a variety of reasons. Operators who miss work and do not provide any forthcoming information as to why they did not show up is not acceptable, at least in a pay-for-skill program. Although a person can progress with this upfront notice, there still will be points taken from the worker's total. Another approach when reviewing the operators' progression is to allow up to a 20% reduction in points, notice or not, and they still can move up. Regardless, create a structured guideline for attendance so the organization is moving operators through the pay-for-skill program when it is truly deserved.

Kaizen and Kaizen Event Participation

Kaizen involves everyone in the company, and the organization should encourage active participation in the continuous improvement initiatives. As the company develops its Lean culture, employees should be allowed to make improvements to their work areas as often as possible. I realize that during the day-to-day operations of a manufacturing floor it's difficult to have operators work on kaizen for three hours; they should be able to generate ideas and develop simple solutions to their everyday work.

I visited a client in 2006 and we had just started our partnership. It was a small family-owned machine shop with gross revenues of about $1 million. The manufacturing floor was about 50,000 square feet and they had been implementing 5S for about a year. It was my second visit with them and I was getting a tour from one of the production supervisors. Their computer numerical control (CNC) machines, lathes, deburring stations, and inspection areas were combined into individual work cells. Each cell did a variety of work while staying within a family of products and like processes.

This particular production supervisor managed cells 5 and 6. We were walking through cell 5 and watching the action. I had a lot of questions for him as this particular cell was schedule for another kaizen event to help decrease setup times and organize the fixture and jig inventory. He was discussing the flow of the cell when he stopped in his tracks and started to look at a desk. This particular desk was used by the cell's line lead who gave the machine operators their schedules and day-to-day tasks. The production supervisor really just managed the line leads.

I myself became curious and asked him what had caught his eye. He pointed to the desk and there was a small cardboard pen holder taped to the desk. The holder was clearly made from a corrugated box that came from a supplier. This makeshift pen holder contained three highlighter pens: green, yellow, and red. The production supervisor explained to me that the holder was not there the previous day and items like that stood out quickly because the company had embraced 5S. Basically, a new item had appeared on a desk that had everything clearly labeled and identified.

We approached the line lead and asked why she had made the box. She told us that every morning she receives the work orders from the office and it is difficult to see what orders had to be worked on first. Some jobs could be done much quicker than others and some had longer delivery dates. The line leads were empowered to distribute work as needed to ensure all orders were done on time while minimizing wait times and imbalances between machines.

After she received the order, the line lead spent about 30 minutes of each morning sorting through the work orders before handing them out to the line workers. The line workers would prepare their work areas as needed from that point. The purpose of the three highlighters was to act as a color code system to quickly identify the order of importance. She simply highlighted rush or fast jobs with a green highlighter. Orders with light completion dates received a yellow mark, and red indicated the least urgent work orders of the day.

The line lead was practicing kaizen and was making improvements to her area. It is this type of mentality that is needed. In regard to the pay-for-skill progression, a company could document these small improvements throughout the plant and keep records of those operators or other floor personnel making the changes. Each level in the program could require a certain number of small implementations to be part of the required progression into a higher paying level.

The company kaizen program includes the encouragement of improvement ideas by using the kaizen event suggestion form. As production workers come up with recommendation for improvements that may require more time and effort, the suggestions then can be considered for a future kaizen event. Ideally, the operator would be asked to participate and make the improvements that he/she came up with. Advancement into another level could simply require production workers to submit a certain number of suggestions.

Probably the most important of the kaizen and kaizen event participation portion of the program is the number of kaizen events in which the operator participated. This part of the pay-for-skill program is a little more difficult because scheduling the events and selecting the teams may differ from one year to the next. If an organization only conducts five kaizen events in a given year, chances are that some production workers may not even get the opportunity to be selected. Ideally, a company should strive to have at least one kaizen event per month; the reality is that the number of events could be lower. It just depends on how aggressively the Lean journey is moving. Once some level of consistency is achieved with kaizen event scheduling, then make it a requirement for progression into the next level in the program

Quality Errors

Production workers are responsible for quality and, with the implementation of quality at the source, an organization can easily track line errors occurring on the line. Operators are human and they are faced with challenges every day that can make it difficult for them to do a job perfectly. Humans make errors, and manufacturing companies need to realize that and give production workers leeway in making mistakes. However, if production workers continue to make mistakes, the process must be analyzed further to help reduce the occurrence. I am a firm believer that the company must effectively design and set up the process so the production workers have all the tools needed to be successful. Once the process is under control, then errors should be at a minimum. If a production worker continues to make mistakes, this should be tracked and reviewed. One of the requirements for advancement in regard to quality could be the number of mistakes made in the process. The goal here is not to point fingers at people, but to once again encourage strong performance and reward them for meeting certain performance standards.

Developing a pay-for-skill program can be very successful for a company. It also can be a headache and create some level of animosity between workers. The whole point of this program is to provide a clear career path for production workers. Production workers, in my opinion, are some of the most valuable employees because they are building products that financially support the organization. Companies usually provide career advancement opportunities for managers and engineers, and little effort is given to providing these opportunities for production workers. Consistent performers and Lean change agents should be rewarded and, I think, a customized pay-for-skill program is a good approach. I hope I have provided some direction and insight into helping you develop one at your facility.

MORE RETURNS

Some companies I have helped took their incentive programs to a higher level. Although pure acknowledgment and appraisal in the long term is the best, you can add yet another monetary incentive program. As a Lean company, you want to continue to encourage production workers to come up with continuous improvement ideas all the time. This is kaizen. As mentioned a couple of times, the kaizen suggestion form is a great way to garner fresh ideas from the production floor, but there is another way to get the workers to make the changes outside of a kaizen event.

These improvements may take some time; they will slowly implement the change as the work days and weeks progress. If an operator sees an opportunity to make an improvement to their process, implements the improvement, and it reaps a financial savings or gain for the organization, award them with a check of some value. For instance, a production worker sees a better way to package the products to reduce time and material. A recommendation can be placed with management and engineers to first see if it is feasible from a product specification perspective. Once the approved changes have been placed into an engineering change request process, the new process change is initiated. After a given period of time, the improvement can be measured to see if it truly made a positive impact on productivity and material cost. Analyze the savings over a year to see the change's annual cost savings. If the improvement saved the company $20,000 annually, as an example, the employee receives a check for $500.

In the beginning, this new program will be slow to encourage workers; but after the first or second idea turns into a payback, more workers will be excited and begin their own improvement projects. These projects, of course, involve other employees, but idea generation is the start. Plus, with a program in place to encourage the behavior, more production workers will become engaged in the process. Each company has to establish some guidelines for the program in regard to annual savings, time frames for implementation, and the amount of money to be paid out to the worker. It is just another approach to soliciting continuous improvement ideas from the factory floor workers.

LEAN TRAINING PROGRAMS

Training new and existing employees in an organization takes on a whole new meaning during a Lean journey. Companies often do not train or invest in their employees, in general, and that can have a negative impact on the company's overall performance. As companies embark on a Lean journey, new and existing employees must be put through some formalized training to ensure that they understand Lean and how it is being applied in their factory.

As Lean implementations occur, and areas of the factory are working under the controlled conditions of 5S and standard work, training is needed to help workers adjust to their new environment. All new employees, including newly hired operators, supervisors, engineers, and managers, must go through a similar process. This chapter is dedicated to the importance of Lean manufacturing training for all employees within an organization. First, I address production workers, and, then, cover managers and engineers.

NEW EMPLOYEE TRAINING PROGRAMS

As your Lean journey moves along, it is important to hire people who can contribute to the continuing success of that journey. Of course, their contribution is contingent upon how well they are trained on your company's Lean initiatives. I think it is wise to introduce Lean to new employees very quickly, and with a lot of excitement, so they get on board right away and realize that Lean is part of the how company operates. Lean is a way of working. Some organizations, when searching for new talent, advertise that they practice Lean in their organization. It can be a good marketing approach to attract the best employees.

Training new employees requires structure. I believe there should be four levels of training, separate from the typical, human resources new employee orientation. How this training is structured and scheduled will vary from one company to the next. I will outline the fundamental aspects of the Lean training plan, and you can adjust it as needed for your company:

Level 1: Company Product Overview
Level 2: Quality Overview
Level 3: Introduction to Lean Manufacturing
Level 4: Mock Line Training

Level 1: Company Product Overview Training

Most organizations have new employee orientation, which involves a variety of topics in regard to human resource functions. Employees fill out employment forms, W-4s, and disclosure agreements, as well as other necessary documentation. Usually, an HR representative presents an overview of the company, the benefits, and the guidelines and policies, which new employees must know in order to work and advance in the organization. After this traditional orientation, employees who will be working in the manufacturing process should attend some form of company product overview training. I call this Level 1 training.

New employees need a fundamental understanding of the products that the organization manufactures and markets. Many times, this vital training step is skipped, and production workers are simply expected to learn about the products while they are on the job. While I believe that hands-on learning is extremely valuable, I also recommend the addition of formalized, product overview training. This training would be conducted in a classroom environment where employees can focus on the information and ask questions.

The product overview training should include real products for employees to touch and analyze. If possible, break down each product into its individual components and discuss the part descriptions. Explain how the product is made and why certain parts are assembled onto other parts. During this training, provide a list of the parts and their part numbers, and train the employees to be able to read the part numbers and understand what the numbers represent. Typically, the individual numbers that make up a part number point to a particular supplier, a stock room location, or a category/family of parts (e.g., hardware, brass parts, wiring, etc.). If possible, provide a list of all the products that are manufactured. Explain which options are available to customers and which parts are associated with these options.

While product overview training can be an ongoing process, I believe it is good practice to provide new employees with some upfront product orientation. The length of this training will vary from one company to another. For example, one particular company created a product overview curriculum that included an initial course, lasting one full day, and subsequent refresher courses that were much shorter and conducted once a week for three weeks. Regardless of the content or length, company product overview training is a smart approach to training new employees.

Level 2: Quality Overview

Quality is critical to company success, and therefore it should be emphasized when new employees are hired. Level 2 of new employee training should be devoted to quality, its importance, and the fundamental aspects of the company's quality program. Chapter 3 described quality at the source, and the importance of placing the responsibility for quality at the point of build. New operators should know that quality is in their hands, and that they will be required to perform certain incoming and outgoing quality checks as part of their daily work.

Discuss the top three, or five, quality issues currently being addressed by the organization. Show them last year's data in regard to customer and service technician complaints. Allow new employees to view internal quality information so that they can identify the common factory errors made before product leaves the building. I don't think this information should be a secret. Awareness is a key factor in getting employees to react and act.

Level 3: Introduction to Lean Manufacturing

Lean manufacturing training, from top to bottom, in the organization is a must. Everyone must be on the same page with regard to how Lean is applied in the company and the direction in which the organization wants Lean to head. It is equally important to show new employees that Lean is *the* way of working at your company, without exception.

Level 3 of new employee training is dedicated to teaching the basics of Lean manufacturing and, more importantly, how it is being applied in your specific environment. After this formalized training is complete, employees will have a better understanding of Lean manufacturing and what they should expect when working within a Lean environment. Five topics should be covered in this training:

1. Eight wastes
2. 5S and the visual workplace
3. Standard work
4. Available time
5. Kaizen

Eight Wastes

The training should provide a description of the eight wastes as well as how they can negatively affect the company. The training must outline and emphasize how working within excessive waste can make day-to-day work life very stressful, especially when workers will be expected to meet specific performance metrics in the areas of productivity, quality, and volume; for example, waiting for the product to arrive in a workstation.

Regardless of their experience, production workers will quickly understand the concept of waste and absorb the information. I believe it's very good practice to teach inexperienced production workers the seven wastes, as they will then be able to identify them very quickly, which is beneficial to the company.

5S and the Visual Workplace

5S training is the most important Lean topic for new production workers to learn. In many cases, 5S is the first major Lean initiative to be implemented within a company. Once implemented into the manufacturing processes on the floor, 5S keeps everything clean and organized, and, I guarantee, the existing operators do not want new employees to join the process without understanding 5S. I have witnessed this before, and it can instantly create animosity toward the new workers. The operators on the line are enjoying the new organization in place and do not want it compromised. Adequate, upfront 5S will make a world of difference in the development of positive relationships between new and existing production workers. If the company is currently in the process of implementing 5S, the training is still valuable, as newly trained operators can make suggestions for improvement in their new assignments.

Standard Work

Standard work is a bit more difficult to explain to new operators. Typically, they will associate the concept with work instructions or some other form of documentation. As mentioned earlier, standard work is the best, most efficient, and safest way of performing work. This definition, alone, may not provide enough insight into what standard work is all about. When describing standard work, it is important to emphasize that there are good and bad approaches to performing all work. Building a product that requires a lot of movement, walking, searching for parts, and waiting on other processes is not a best practice. When most of the work time is spent actually performing the intended work, it is close to performing standard work. Here is a list you can use in your training to help workers understand the concept a bit better.

Standard work is

- Work content in a workstation
- Quality responsibilities in a workstation
- Inspection procedures
- Testing procedures
- Safety instructions
- Setting up a machine
- Material handler routes and routines

- End-of-day cleanup procedures
- Work instructions

These are all good examples of standard work, and there is a best or recommended way to perform all of these.

Standard work is not

- Leaving the workstation to find parts
- Leaving the workstation to find tools
- Sorting through piles of work instructions
- Waiting on information
- Taking extra breaks
- Talking on cell phones

Standard work is about defining clear roles and responsibilities throughout the factory with minimal or no waste.

Available Time

It is critical that this topic be part of the Lean training curriculum. New operators joining the company are not likely to have heard of this concept. It is important to explain that Lean manufacturing processes are designed to maximize work time. The importance of working together and leaving for breaks and lunch as a team should be stressed. Explain that workers will be held accountable for specific volume requirements; therefore, utilizing their "touch time" is critical to daily success. Remember: you are not operating a labor camp, so be diplomatic in your approach. The key is working smarter, not faster.

Kaizen

The concept of continuous improvement will become a way of life at your company. Lean and kaizen is the desired approach to running your business. Therefore, you need to take advantage of the fact that these workers are new and have not yet developed any bad work habits. They left those bad habits with their previous employer. The final component of your Lean training is to describe the concept of kaizen and how workers will be involved in kaizen events throughout the year. It is wise to let them know early on that they will be asked to make improvements to the organization.

Teaching the theory of kaizen is challenging. Explain that the kaizen events will be executed often and that workers will be asked to participate in these projects. Discuss the purpose of the kaizen event suggestion box, and go over the kaizen event suggestion form and how to complete it properly. Show them the box and discuss how the suggested ideas will be considered in order to make improvements to the company. Display pictures of the prior kaizen events, if you held any, and show them a copy of the kaizen newsletter. Walk the workers around the production floor, identifying the location of the kaizen communication boards and discussing their purpose. Kaizen and Lean is serious business, so it is an important concept to be included in initial training.

Level 4: Mock Line Training

Manufacturing is a dynamic environment, completely different from other industries. Some of the new production workers who apply to work at your company may not have manufacturing backgrounds and may be accustomed to different working styles. For example, workers in retail often have times when customer flow is slow. Many workers tend to prepare for the "rush," work the rush, and then recover from the rush. In some jobs, the workload levels vary from one day to the next. It's important to realize the distinction between other environments and manufacturing, which is highly aggressive and fast-paced. Not to down play the service industry, at all, as they have their own busy times. But, the fact is, manufacturing and service are completely different environments.

Service workers bring great knowledge and, usually, a hard work ethic. However, they must be adequately prepared for manufacturing work. Level 4 of new employee training involves working out on the production floor, away from the real manufacturing lines, in what I like to call "mock line" training.

A mock line is a training area in the factory that looks, and works, like real production. It is important to design and construct this area to be identical to the manufacturing process for which workers will eventually be responsible. Your mock line training area should be set up based on the processes in the plant. More importantly, it must contain all the necessary Lean programs you have in place.

Identify an actual assembly process in the plant; something small, that can be easily copied and constructed. Create enough workstations in the mock line to allow the new operators to experience working side by side. Each workstation should contain all the necessary tools to build the prospective product: air tools, parts, bins, fixtures, hardware, work instructions, and safety equipment. This mock line should have all the 5S detail in regard to floor tape, station signs, designations, tower lights, and a material replenishment system. The new operators should be taught how to build real products, working within a Lean process that is controlled by structure and protocol. I realize that not all manufacturing processes utilize assembly-based systems. Therefore, if your company is either a job shop or a fabrication-style business, set up your mock line as appropriate for your specific environment. For example, a composite factory, like fiberglass, should establish a mock line that teaches how to do hand lay-up.

New operators can learn the concepts of single-piece flow or controlled batches. They can be taught and then shown how flexing from one workstation to another is accomplished. This type of training environment allows them to learn how to work in a Lean process, and be prepared to go to work once they are assigned to a real production line. Existing operators will appreciate the mock line to ensure that new operators follow the rules of the Lean process. It will shorten the learning curve of new operators and allow them to build a quality product for the organization.

The four levels of new employee training will take some time to set up, and to put all employees through the curriculum. Mock line training is likely to take at least five days to complete, which may be too long, or too short, depending on each manufacturing environment and how the company wants to conduct the training. It is possible that a company may not be able to afford the recommended four levels of training for new employees. Decisions will need to be made. But, regardless of the company restrictions, training new operators is extremely important to the success of the Lean implementation and the company as a whole. Consider it something you can't live without.

New management and engineers often receive formal training when they join a company. Even manufacturing companies tend to invest more heavily in support staff training than on the training for production workers. I believe that adequate training is essential for all employees, regardless of their position or title, as it reflects on the success of the company. I have heard management's argument—believing that production workers will just take their training to another company where they will be paid one dollar more per hour. Simply put, management believes that investing in training for line workers will only benefit someone else. That being said, it is bad practice to automatically discard training because of the fear of turnover. There are plenty of employees who will stay with the company and their training would not be a waste of money. Invest in your people and you will reap the benefits.

CROSS-TRAINING PROGRAM

At this point, you have some well-trained employees working in your processes. They will take some time to learn their new environment, but at some point you will want them to learn more stations, operations, and other processes. Flexible production operators add tremendous value to any company, especially in a Lean workplace. Operators must be cross-trained, over time, to learn more jobs and operations. This allows production supervisors to assess the skill level of their people and be able to shift workers to different areas of the floor as needed. All existing employee training curriculums need to have a cross-training component. Cross-training must be customized to fit each company and each process within the company. Here are three considerations for your cross-training program:

1. Levels of progression
2. Temporary worker progression
3. Cross-training matrix

Levels of Progression

Before creating a cross-training program, levels of progression must be established, identifying the specific job skills an employee must have in order to move up to another level within the company. Three categories should be established: entry level novice (N), certified (C), and trainer (T).

Novice (N)

An operator at the novice level is a worker who is new to the workstation, or has just been hired. Essentially, this person is an entry-level employee who has just attended the four levels of training, or an existing employee who has transferred from another area of the company. Either way, this employee has passed the initial training process. Depending on the company, new employees may have a 90-day probation period, allowing the employer to assess how they handle the work. I think this is also a good policy for existing employees being trained in a new workstation or area, perhaps with a shorter probationary period, of 7 to 10 days.

Novice operators are given some leeway and should be allowed to make minor mistakes while they are learning. Place the novice operators with experienced workers so they can be helped if necessary. Determine the amount of time that experienced workers will assist novices, and then let them work on their own. Monitor the novice's ability to follow the work instructions, perform the required work, adhere to cycle times, and conduct quality checks. Learning curves will vary, depending on the complexity of the product, the number of tools, and the abilities of the new operator. Once they have worked to the end of the probation period, you will make an initial decision regarding whether they should be considered for certification status, the next level. Many factors will influence your decision, such as if they are new or existing employees, as well as their performance during training. Nevertheless, they should remain at the novice level until they have had the opportunity to be consistent performers for a specified period of time, which you will decide. Novices must work within takt time, perform the work content, and conduct quality checks. They also must be able to react properly to resolve any issues that arise. The production supervisors should conduct a few audits during this period of time to see how the new workers are doing. Once the workers have proved that they can work consistently, day in and day out, for the given time period, without supervision, then they are ready for certification.

Certified (C)

The second level of the cross-training program is becoming certified. Certified operators "run the show." Workers who are certified may remain at this level for a long period of time. Although they have the experience and knowledge to "run the show," and are capable of making good decisions regarding quality errors and flow, they may not be ready to provide training to others. Training others is almost an art form. It takes time and practice to become comfortable dealing with people and teaching them what you know. Many people really understand a subject or know how to perform a job, but cannot relay that information to others in a way that is clear and easily understood.

Working in a Lean process takes skill, and that skill, is developed over time. Certified operators understand their responsibilities in their own workstations as well as the procedures and protocol that control those processes. So, when are they ready to train others? Naturally, certified operators should strive to reach this next level of job performance on the floor. However, I believe that it is wise to offer the option of moving onto another workstation, or area, first, before progressing to the training level. Certified operators will become novices in the new area or workstation and have the opportunity to cross-train into another skill or skill set. Some operators will see the value in that offer and choose that path. The cross-training option is wise because some certified operators may not feel they are ready to train others, and may want to continue progressing through the various workstations or areas first. Others will want to move directly toward a training certification status. Of course, management should always persuade workers to move in the direction that is best for the organization as a whole; but giving the operators the choice empowers them to be in more control of their own advancement, which is very positive.

Both choices outlined above are natural progressions for a certified operator. Each choice has its own requirements. If the certified operators decide to try their hand at a new workstation, they will follow the novice-level progression guidelines, as mentioned earlier. To become a trainer in a particular workstation, the certified operator must work in that station for a specified period of time, possibly three to six months. The exact amount of time really depends on your company's manufacturing processes. I also recommend a series of tests to be performed on the production line in order to verify the operator's conformance to standards. Some certified or veteran operators may begin to ignore procedures and lapse into poor work habits, which is just human nature and happens with salaried support staff as well. Creating a testing and auditing system for certified operators who would like to achieve trainer status is the best method of qualifying them for advancement.

Trainer (T)

Workstation trainers have mastered their area. Once employees have successfully gone through the new employee training program, they are placed at a workstation with another operator who has trainer status. The goal of any cross-training program is to get all permanent employees to achieve trainer status.

Temporary Worker Progression

Temporary workers exist is just about any industry. Manufacturing is not an exception, and more organizations are using temporary workers to help during times of increased output. Certain times of the year are busier than others, depending on the products and industry needs. For example, companies that manufacturer gas stoves for heating are busier from late summer through winter. Construction products like vinyl windows and siding are manufactured heavily during spring and summer. Rather than having a larger, permanent workforce all year round, companies simply bring in temporary workers during the busy seasons.

This approach to employment also is used to identify good, reliable workers who can eventually become full-time permanent employees for the organization. Because temporary workers often become permanent, a system of progression should exist for those employees, identifying the time periods spent in various activities and the point that they will be eligible for permanent status.

Many companies do not develop such a progression, and temporary workers remain temporary for years. I firmly believe in developing a solid pool of talent, and temporary workers are a wonderful source from which to pull this talent. Any company embarking upon a Lean journey should be proactive, identifying early on those workers who have the attitude and ability to assume future responsibilities. The temporary worker pool is a great source for finding that talent.

New talent always improves the whole pool. Sometimes, the workers who have been with the company the longest and have the most knowledge and experience are the most resistant to change. New workers bring a fresh attitude and approach, and have not yet established poor work habits. The blending of these workers is beneficial to your organization, as they can balance each other, and inspire and motivate one another. Both types of workers are necessary for successful Lean implementation.

Temporary workers are usually seeking permanent employment. They come with new ideas and attitudes, and are eager to learn and to prove their worth. Most of them want to become part of your company. Yes, there are also those who simply come for a time and then leave. However, in most cases, temporary workers really want to become permanent and will do their best for you. I have seen temporary workers who can outperform many certified operators. I have even witnessed temporary workers training permanent workers.

TRAINING MANAGERS AND ENGINEERS

New employees entering into support staff roles also need training that sets correct expectations for work within a Lean organization. I think these individuals should go through the same training curriculum as production workers, in regard to product overview, quality, Lean, and mock line training. Are you surprised? Many Japanese manufacturers require their new managers and engineers to work on the production floor for a given period of time. It allows them to see how the products they will be supporting are made, and also encourages the development of working relationships with the operators. Working on production lines also provides insight into issues that create obstacles for operators. While I am not necessarily recommending this approach, it is certainly something to think about.

Managers

As new managers come into the organization, they should be briefed on the company's Lean journey, preferably during the interview process. Some companies make Lean experience a requirement and list it in the manager job description. The kaizen steering committee is an integral part of the Lean program, and new managers may become part of this decision-making group. Employees from their departments will be asked to participate in kaizen events. Managers for whom the concepts of kaizen and Lean are foreign will not be prepared for what takes place, if not properly informed early in the hiring process. New managers should be aware that their input, contributions, and involvement are required for the continued success of the Lean journey.

Engineers

New engineers also play an important role in the Lean journey. Manufacturing, industrial, and process engineers will assist with data collection for time studies, waste analysis, process, and value stream

mapping. These individuals should go through some form of waste identification and standard work training to show them the tools needed to
perform the technical side of Lean manufacturing. If the organization has
hired a full-time Lean/kaizen champion, new engineers should be made
aware of this person because they will help each other capture the current state of individual processes. Plus, at some point, the current kaizen
champion may leave the process, or the company, and new engineers must
be prepared to take over the role and responsibilities.

As with job descriptions for new managers, the descriptions for new
engineers should reference the company's Lean journey, stating the
importance. Possibly, new hires will already be Lean familiar and perfect candidates to step into the new role. Although not a requirement, this
would certainly shorten the time needed for training. New engineers also
should be informed that they will be asked to participate in kaizen events
and perhaps even take on a leadership role.

Keeping your Lean journey going through different CEOs and owners may be challenging as some people may want to completely discard
everything. I have come across professionals who actually found Lean to
be a poor business practice. It was my experience after talking with them
that the companies they worked for prior, implemented Lean and practiced it incorrectly.

There is no telling what a new company leaders will do once they start
working, but I do recommend developing programs and systems to continually train your people and provide a foundation that harnesses continuous improvement.

Conclusion

Humans react to the environment around them and as you change that environment, so will their perception of it. Poorly run companies with inefficient processes will have a negative impact on the employees. Lean is a powerful improvement tool that not only can change company performance, but also positively change how the employees feel about working. Waste-ridden processes only create dysfunction, provide no sense of targets, institute failure, and reduce forward-thinking approaches to working. Lean is not the one and only business model. It must be incorporated into your existing business model to further assist in achieving critical business goals. Your journey will be different from everyone else's and often a Lean best practice in one place may not work in yours.

Academic-type leaders like nice step-by-step guidelines. There are no Lean textbooks. As you implement what you have learned in this book and what you learn from other sources, you will find how it applies to your company. It works. When rolled out properly, Lean works. More importantly, it creates a result-driven culture and, as you move from post to post into other companies, I want you to look back and see that you have been part of changing a culture first and business metrics second.

Glossary

3 Main Drivers in a Business: Cost, Quality, Delivery. Companies must operate under conditions that offer a competitive balance between these three main drivers. It is difficult to accomplish as each customer has different needs in regard to these drivers. Application of Lean manufacturing principles can help in getting close to an optimal balance of cost, quality, and delivery.

5S: A methodology for organizing, cleaning, developing, and sustaining a productive work environment.

    Sort: The act of removing all unnecessary items for the work area.

    Set in Order: The act of organizing what is needed so it is easily identifiable in a designated place.

    Scrub/Shine: Clean everything.

    Standardize: Consistency and best practice.

    Sustain: Maintain the organization and provide opportunities for improvement.

5S Audit Form: This form is created and used to help monitor the 5S program and to ensure that improvements are made when there are deviations from the program. Audits should be performed on a weekly basis to ensure the program is sustained.

5S Supply Box: A box used to house all the items necessary to support a kaizen 5S event.

5S Tracking Sheet: A visual tool that is displayed on the production floor to show how each process or area is sustaining 5S. The scores come from the weekly 5S, audits and the tracking sheet is updated once a month. Incentives should be provided to the process or area with the highest monthly score.

8 Wastes: Overproduction, Overprocessing, Transportation, Waiting, Motion, Inventory, Defects, Wasted Human Potential.

Controlled Batches: An inventory and volume control approach to ensuring the right amount of product and parts is being built when needed. Operators must build to the set controlled batch quantity, then stop and verify the quantity and quality.

Cross-Training Matrix: A management tool used to train and monitor an operator's skill level within an assembly line or other manufacturing process. There are three skill levels: trained, certified, and certified to train.

Current State: A document showing the current state of a process.

Defects: One of the Eight Wastes; defects are often hidden in excessive WIP (work-in-process). Caused by overproduction, insufficient training, or inaccurate or outdated standards/instructions.

Effective Hours: Effective hours represent the amount of time production workers actually spend building products and fabricating parts. It excludes morning meetings, breaks, lunches, end of cleanup, and any other scheduled time away from the manufacturing process.

External Work: Setup activities performed when the machine is up.

In-Process Kanban: A visual signal that pulls material or information from the previous step in a process. A WIP (work-in-process) management tool.

Internal Work: Setup activities performed when the machine is down.

Inventory: One of the Eight Wastes; excessive inventory takes up floor space and creates a longer-than-needed manufacturing process. Excess inventory ties up money in WIP that can be used for other purposes.

Kaizen: Japanese word for continuous improvement that encompasses the idea of employee participation and promoting a process-oriented culture.

Kaizen Champion: An employee who is 100% dedicated to kaizen and driving the continuous improvement efforts within the organization.

Kaizen Event: A preplanned, scheduled, process improvement project intended to implement Lean manufacturing principles. Kaizen events are planned four weeks in advance to ensure 100% participation of team members and achievement of the event goals.

Kaizen Event Tracking Sheet: A spreadsheet that is used by the kaizen steering committee to plan and track all kaizen events in the organization to ensure completion of all kaizen-related activities.

Kaizen Steering Committee: A group of upper managers who oversee all kaizen event activities in the company. The committee meets once a month and is led by the kaizen champion.

Kaizen Suggestion Box: Used for collecting employee recommendations on continuous improvement ideas.

Kaizen Team: A group selected to participate as a team during a kaizen event.

Kanban: A concept of "signal" to simplify communication. These signals are used for material replenishment on production floor, maintenance, and in support functions.

Kanban Card: A "signal" card used to communicate the need to reorder an item. It includes part identifier, reorder quantity, and both production and inventory location.

Lean Manufacturing: A methodology of the removal of inefficiencies called *waste*. It is a business philosophy of continuous improvement to better satisfy the customer and to become more competitive.

Lean Metrics: Productivity, floor space, travel distance, inventory, resources, quality.

Line Balancing: An approach to shifting work around a production line to keep work even among workers.

Mistake Proofing: The use of any automatic device or method that makes it impossible for an error to occur.

Motion: One of the eight wastes, when operators leave their workstations, unnecessary reaching, walking to and from maintenance.

Nonvalue-Added: Process steps a customer is not willing to pay for.

One-Turn Method: A concept of implementing quick disconnect and reconnect devices on machines.

Overall Equipment Effectiveness (OEE): These calculations report how well production is performing based upon three metrics: availability, performance, and quality. Total Effective Equipment Performance considers the impact of loading on OEE.

Overprocessing: One of the Eight Wastes, overprocessing is redundant effort. Common when the end of a process is undefined (e.g., sanding, polishing, or deburring).

Overproduction: One of the Eight Wastes, overproduction is producing more than needed, before it is needed, and faster than necessary.

Pay-for-Skill Program: An incentive-based program to encourage multiskilled workers. As production workers learn more jobs, there is a pay increase for becoming proficient in more areas of the company.

Poka Yoke: A Japanese word that means mistake proofing. This is a way to prevent mistakes and defects in a process addressing human-caused error.

Production Control Board: A visual schedule board that shows real-time information on production, often by the unit or hour.

Red Tag: During 5S events, a red tag is attached to any item sorted out as not needed.

Red Tag Area: An area set aside for temporary storage of red tagged items.

Red Tag Register: A list itemizing every item red tagged during a 5S event.

Red Tag Removal Procedure: A defined procedure for handling items given a red tag during a 5S event.

Rejects: One of the Eight Wastes; also referred to as scrap, which suggests that the time and materials taken to make a reject is wasted.

Reorder Point: The point within the kanban system where the signal indicates it is time to reorder that particular item.

Right Sizing: Concept of space utilization; utilizing only the amount of horizontal or vertical space required to store an item.

Single Minute Exchange of Dies (SMED): Lean manufacturing concept of quick changeover that accomplishes setup in nine minutes or less.

Single-Piece Flow: The movement of parts or units one piece at a time in a manufacturing process.

Sorting: 5S term meaning capturing the removal of all unnecessary items from a work area.

Standard Work: An agreed-upon set of work procedures that establishes the best and most reliable methods and steps for each process and each employee. These methods are clearly defined, represent best practice, and are supported by documentation.

Standard Work Combination Sheet: Data collection sheet used to collect work steps and cycle times.

Standard Work Procedure: A detailed, visual work instruction sheet outlining the agreed-upon best practice within a process. The Standard Work Procedure is posted in the work area for operators, and includes pictures, icons, but very few words, allowing for the prevention of language barriers commonly found in the workplace.

Strategic Purpose: A Lean manufacturing business strategy used as a guideline for implementations and improvements. The strategic purpose is revised once a year.

Sustain: A 5S term suggesting the need and methods to maintain the new state of the process standard.

Takt Time: German word for rhythm, which is the time that a unit must move from one workstation to the next to meet the required daily output. It represents the product completion interval for a given process.

Total Productive Maintenance: Company-wide approach to increase the effectiveness and reliability of equipment.

Tower Lights/Andon Lights: Colored light system that is installed at the workstation to allow operators to communicate with support staff and material handlers.

Transportation: One of the eight wastes, it is the physical movement of information or material. Often caused by overproduction, poor planning and scheduling, and inefficient plant layouts.

Two-Bin System: A replenishment system where parts are pulled from one bin only after a second bin has been emptied; the empty bin is the signal to have it refilled.

Value-Added: Process steps a customer is willing to pay for.

Value Stream Mapping: The process of mapping out each step in a facility—from the customer order to shipment of the product. The Value Stream Mapping process uses symbols and time studies to produce a map of the facility. The map then can be used to evaluate waste reduction (or avoidance) opportunities. The map shows both the flow of information and product.

Waiting: One of the Eight Wastes; when manufacturing processes are out of synchronization causing operators and machines to be idle.

Wasted Human Potential: The Eighth Waste, which underutilizes people's skill sets. People are only as effective as the process they are given to work in. A process that has the other seven wastes automatically creates the Eighth Waste.

Index